DATE DUE

Gold,
Money
and the Law

Gold, Money and the Law

edited by
HENRY G. MANNE
ROGER LEROY MILLER
University of Miami

for the
Center for Studies in Law and Economics
University of Miami School of Law

Aldine Publishing Company, Chicago

About the Editors

Henry G. Manne received his undergraduate training in Economics at Vanderbilt University. Professor Manne received his JD degree from the University of Chicago Law School in 1952 and in 1966 received a JSD degree from Yale Law School. He is a member of the Illinois, New York and U.S. Supreme Court bars and serves as Chairman of the Association of American Law Schools' Section on Economic Regulation. Professor Manne has taught at St. Louis University Law School and George Washington University Law School, and until 1974 was Kenan Professor of Law and Political Science at the University of Rochester. He has contributed numerous articles to the literature and is the author of *Economics and Legal Relationship* and co-author with Ezra Solomon of *Wall Street in Transition*. Dr. Manne is currently Distinguished Professor of Law and Director of the Center for Studies in Law and Economics at the University of Miami School of Law.

Roger LeRoy Miller received his undergraduate training in Economics at the University of California, Berkeley, and his Ph.D. in Economics from the University of Chicago in 1968. Dr. Miller has taught at the University of Washington and has served as an economic advisor to both governmental and private agencies and institutions. He has contributed numerous articles to the literature and has published some 20 books including *Economics Today, Applied Econometrics* (with P. Rao), *The Economics of Energy,* and *American Economic Life.* Dr. Miller is presently Professor of Economics and Associate Director, Center for Studies in Law and Economics, University of Miami School of Law.

Copyright © 1975 by The Center for Studies in Law and Economics, University of Miami School of Law

First published 1975 by
Aldine Publishing Company
529 South Wabash Avenue
Chicago, Illinois 60605

ISBN 0–202–06072–1 clothbound edition
Library of Congress Catalog Number 75-20705
Printed in the United States of America

*Dedicated to the memory
of Pierre F. Goodrich*

Contents

Introduction
  Henry G. Manne and Roger LeRoy Miller   1

Gold, Money and the Law: The Limits of Governmental
  Monetary Authority
  James M. Buchanan and T. Nicolaus Tideman   9

Gold, Money and the Law: Comments
  Milton Friedman 71
  Harry G. Johnson 83
  Ralph K. Winter, Jr. 93
  Gerald T. Dunne 107

Response to Comments
  T. Nicolaus Tideman 119
  James M. Buchanan 125

Conference Participants 131

Discussion 135

Index 211

Introduction

HENRY G. MANNE
ROGER LEROY MILLER

The Center for Studies in Law and Economics was established at the University of Miami to bring together scholars in the areas of economics and law. This appeared to be a most propitious time for the creation of such a Center. Economists are now realizing that many, if not most, legal issues have serious economic implications; and, perhaps to a lesser extent, legal scholars are coming to realize the importance of economic analysis in legal research.

For the Center's inaugural conference, we sought a significant legal topic that had been overlooked by economists as an area where their expertise could be usefully applied. There was much talk in early 1974 about the price of gold and the belief that Americans could not own it. No one, however, was considering the impact of the legal arrangements resulting from the decisions taken by the federal government in 1933 prohibiting American citizens from owning gold.

More specifically, few, if any, economists had ever addressed themselves to the Supreme Court cases in 1935 upholding Congressional abrogation of existing and future gold clauses in either public or private contracts. It therefore seemed that these classic *Gold-Clause Cases,* nearing their fortieth anniversary, would be a highly appropriate topic for the Center's first interdisciplinary conference bringing together law professors and economists. The papers and edited discussion from that Conference, which make up this book, demonstrate how law and economics can be integrated in grappling with major legal and economic issues.

* * * * * *

There had been much talk by the time of the Conference in late 1974 about indexing, or adding escalator clauses to contracts, so that private citizens could be protected from widely varying rates of inflation. Indexing, of course, is not a new concept. Some students of the subject trace its origins back many hundreds of years to contracts made in England that were tied to the price of specific staple commodities. Alfred Marshall made reference, on several occasions, to what he called the "tabular standard." Other economists, both in the United States and abroad, have talked in similar terms about indexation.

In the history of the United States, a form of indexation became widely used after the era of "greenback inflation." Paper banknotes were issued by the Union in order to finance the Civil War. These "greenbacks" had declined in value by about one-third (compared to gold) by the end of the Civil War. In response to such rapid declines in purchasing power, private citizens began to insert gold clauses in their contracts. Such clauses called for payment in gold, gold coin, or its equivalent. In other words, these contracts were tied to the value of gold, which historically appeared to be more stable than the value of banknotes. The Supreme Court, during the 1880s, held that a contract which called for payment in gold dollars was, indeed, enforceable.

Gold clauses flourished until the 1930s. A typical gold clause, such as that added to the bonds of the Baltimore & Ohio Railroad, issued on February 1, 1930, read that payment of principal and interest would "be made . . . in gold coin of the United States of America of or equal to the standard of weight and fineness existing on February 1, 1930." Gold clauses of the type just mentioned became a routine provision in many debt and sales contracts, including contracts on which the government itself was the debtor.

All this was to change in 1933. A joint resolution of Congress on June 5 of that year declared that gold clauses were "against public policy; and no such provisions [could] be contained in or made with respect to any obligation hereafter incurred." Such a resolution followed in the wake of the famous bank "holiday" and the confiscation of all private gold, gold certificates, and some gold coin. Then, the dollar price of gold, for the purposes of international trading, was raised, thus raising the value of gold.

A series of gold-clause cases then went through the courts. After all, the dollar had devalued 40 percent as compared to gold. Debtors were thus able to pay back creditors at a 40 percent discount, given that gold clauses were now outlawed and that obligations could be discharged by payment of "any coin or currency which at the time of payment is legal tender for public and private debts." The principal *Gold-Clause Case*, testing Congressional authority under the Constitution, was *Norman vs. Baltimore & Ohio Railroad Company*. It involved a minor amount—$22.50. Mr. Norman wanted his interest coupon on a B&O bond to be paid back in terms of gold, which at that time would have been $38.10. The court ruled against Norman in a five-to-four decision. Thus, the joint resolution of Congress was upheld, and its constitutionality reaffirmed.

* * * * * *

The principal paper at this conference was presented by Professors James M. Buchanan and T. Nicolaus Tideman. They examined the economic and legal aspects of the abrogation of gold clauses, particularly in terms of contractarian analysis and our implicit monetary constitution. Comment papers by Professors Milton Friedman and Harry Johnson explored both the historical setting in which the *Gold-Clause Cases* were decided and the role of gold in the monetary system at that time. Professor Ralph K. Winter examined principally the constitutional issues that were under consideration at the time. And, finally, Professor Gerald T. Dunne presented a monetarist view of the gold clause question and concluded with a proposed "Anti-Inflation Act of 1975."

Following the papers is an edited transcript of the general discussion that occurred among those attending the Conference. This group included some of the leading economic and legal authorities in the country.

The attendees first discussed the economic history underlying the *Gold Clause Cases* and the role of gold and money after 1933. The relationship of economics to law prior to and after 1933 was the next topic of discussion, with specific emphasis on the legal position of gold after December 31, 1974. Then the indexing power of gold clauses was examined, as well as other ways to index contracts. Finally, the monetary constitution and methodological considerations were examined.

One issue, only obliquely touched upon, concerns the effect of abrogation of gold clauses by Congress on our constitutional system. Had the Supreme Court in these cases permitted legislative abrogation of private agreements and in effect denied private contractual defenses against inflation? Perhaps in the context of government control over numerous economic activities, abrogation represents no significant constitutional change. Indeed, many current holdings strike down contracts because courts perceive an "unequal bargaining strength of one of the parties." If courts can rewrite contracts to reflect their own ideas of what we might call social justice, then the abrogation of gold clauses would seem to follow the same line.

At one point in the general discussion, the doctrine of frustration of contract purpose was argued to justify abrogation, because gold clauses presumably designed to protect against the impact of inflation arguably created an opportunity for a double gain in a time of depression. But that argument leads to an important constitutional consideration: Was the Congress the appropriate body to apply such a doctrine to all gold clauses?

It could certainly be strongly argued, as it was, that, if we wish to keep faith with traditional and current thinking about the separation of powers between the courts and the legislature, a legislature was not the appropriate deciding body. A legislature, by its very nature, can have no knowledge of the intent of parties to individual private contracts, the very kind of factual determination courts are specially suited to decide. Along these same lines, it was argued that the legislature may have overstepped its bounds by not merely abrogating gold clauses then existing but by outlawing them for any period in the future. It was thus suggested, though not strongly endorsed by the lawyers present, that this might even be an assumption of judicial power by the legislature in violation of Article III of the U.S. Constitution.

As could be expected, the question quickly arose about what would happen to the value of gold after it became legal for private citizens of the United States to own it. The discussion of this point, a highlight of the Conference, must certainly have dismayed some advocates of a stronger role for gold in U.S. monetary affairs. Gold had appreciated dramatically in the previous few years and thus would have served, and did serve, as a strong hedge against inflation. However, private citizens in the United States could actually own gold legally prior to December 31, 1974, simply by purchasing numismatic coins. Austria had been minting new gold coins with old dates that had been selling in the United States for several years, and gold in the form of these coins could be bought by paying an extremely small premium.

Thus, American citizens, at no great expense, could own all the gold they wanted. Given that understanding, it was not surprising that a number of the participants predicted that very little would happen to the price of gold after legalization, and, indeed, very little has—though other forms of government intervention in the market may have had some effect.

But several speakers argued that predicting the price of gold is no different than predicting the price of any other asset. In a competitive market, in the absence of inside information, prediction is about as good as random selection. There is literally no difference between predicting the price of a stock on the New York Stock Exchange and predicting the price of gold in the world market. All of the theories related to the random-walk hypothesis for stock pricing hold equally for gold. All current information about the demand and supply (including future demand and supply) of gold is discounted into the current price. Thus, the argument was made that prior to December 31 the world price of gold fully discounted all available knowledge about any increased demand for gold that would occur on the part of liberated Americans.

Pursuing the argument that gold is no different than any other asset, some participants wondered about the usefulness of gold clauses to index contracts. If, indeed, the purpose of a gold clause was to protect both parties (but more importantly, perhaps, the creditor) from the ravages of inflation, then why would not a pure escalator clause be the most appropriate instrument? The Consumer Price Index, it was urged, must be a more reliable indicator of the rate of inflation, on average, than the price of gold. It was therefore concluded by some discussants that, even though private holdings of gold were to be legalized, and even though it might become possible to use gold clauses in contracts, it was not at all clear that such clauses would, in fact, become common again.

As we write this introduction, much of the economic discussion summarized above, about what would happen with gold in 1975, has proven to be correct. However, many constitutional issues still remain; they concern the government's power to abrogate private contracts mutually entered into by fully consenting, adult private citizens. These constitutional issues will no doubt be with us for many years to come.

Gold, Money and the Law: The Limits of Governmental Monetary Authority

JAMES M. BUCHANAN
T. NICOLAUS TIDEMAN

Center for Study of Public Choice
Virginia Polytechnic Institute
and State University

I. Introduction

Effective in January 1975, American citizens can own and exchange gold in any form, something that they have not been able to do for almost forty two years. These four decades commenced with a series of events that raised many issues of "law and money," events that aroused a flurry of discussion among lawyers and economists before fading away in a general acquiescence to the consitutional revolution that the New Deal implemented. Significant modifications were made in the basic monetary structure of the United States in the 1930s, and these deserve reconsideration. Specifically, the government's initial action in forbidding the private ownership of and trading in gold and in abrogating the gold clauses

We are indebted to our colleagues, Gordon Tullock and Warren Weber, for helpful comments on early drafts of this paper. We are also indebted to our formal critics, and especially to Professor Milton Friedman, for detailed comments on the draft circulated for the conference.

in contracts is worthy of modern analysis and interpretation. By concentrating on the actions through which the government swept away these protective devices, we can discuss a defined historical sequence and, at the same time, examine still relevant issues that involve the fundamental limits of government's monetary authority.

"Law and money" offers one of the most challenging interfaces between law and economics, an interface that is worth exploring in the generalized revival of interest in legal constraints on economic behavior. The challenge here is enhanced by the demonstrable failures of modern governments to control inflation, failures that prompt searches for means of adjusting to this acknowledged feature of modern economic life. What legally viable means of protecting and preserving real value are available to individual citizens? Cost-of-living wage contracts, escalator clauses, index-clause bonds and notes, indexed income tax brackets, index contracts generally—these protective instruments generate controversies once again over the whole set of issues concerning the legal boundaries of governmental monetary authority. The constitutional limits of governmental monetary power in relation to the institutional instruments designed to protect persons against abuses of this power must be reexamined.

To avoid misunderstanding, it will be useful to set out our own conceptual framework. We are not legal positivists, historical determinists, or strict constitutional constructionists. On the other hand, we make no claim to a superior wisdom that makes our version of the "good society" more deserving of attention than the next man's (if indeed we could agree between ourselves on such a version). This leaves us with the contractarian framework, the only conceptual schema that is consistent with an individualistic-democratic value structure. This implies that normative justification for potentially coercive governmental actions, past or present, is located only in putative contractual agreement that might emerge from a rational decision calculus of free individuals, treated as equal participants in some constitutional stage of deliberation.

We shall use this framework to evaluate the changes in the monetary order of the United States that were made in the 1930s. We shall, directly or indirectly, raise the following questions. What is the role of the State in monetary matters? What monetary agreements might emerge from genuine social contract in which persons choose the constitutional rules under which they are to carry out their everyday economic interchanges? How did the monetary rules and institutions in existence before 1933 match up to such contractually-derived criteria? Had the United States government, before 1933, met its own constitutional obligations? What might citizens have reasonably expected from government in the 1930s? How and to what extent were reasonable expectations violated by the restrictions on gold dealings and by gold-clause abrogations? What should have been the implications, legal and economic, of these governmental actions? Finally, what is the relevance of that period's history for problems in the 1970s and beyond?

These are among the questions that we shall try to answer in this paper. Section II provides a brief historical sketch of the monetary system prior to the Great Depression, along with the necessary analysis for understanding the vulnerability of this system and the situation that confronted the government in 1933. Section III discusses the policy measures actually taken by the New Deal. Section IV concentrates on the issues of constitutional law raised by the set of governmental actions, and notably, the abrogation of the gold clause in contracts, public and private, and the judicial confirmation of this action. Section V shifts to a discussion of the overall structure of a contractually-derived monetary constitution, and here we look at the observed historical record in the light of contractual criteria. Section VI continues the discussion with attention to the alternative monetary arrangements that might emerge from genuine social or constitutional contract. Section VII represents a necessary detour into the realm of welfare economics to examine the grounds for possible governmental intrusion into voluntary private contracts among persons. Section VIII looks at the record of past failures on the part of government and at the implications of these events for constitutional interpretations of monetary powers in the post-Depression period. Section IX applies the whole analysis to the situation facing us in 1975, trapped in continuing and apparently uncontrollable inflation. We discuss the relevance of the interpretation of the monetary constitution to indexation of contracts in the modern economy. Section X offers summary conclusions.

II. Before the New Deal

Permanent modifications were made in the monetary rules and institutions in the 1930s; the monetary consitution was changed, with effects that continue to be felt forty years later. The nation experienced economic disaster in the 1930s. Understanding this, much can be excused. The confusion and contradiction among separate policy instruments, the haste with which actions were taken, the apparent disregard for long-run consequences, the failure of orderly political process—these deserve sympathetic explanation.[1] An appreciation of the setting for the policy choices of the 1930s does not, however, absolve the modern scholar from his duty of examining these institutional changes independently. We

---

1. For an *un*sympathetic but interesting summary description of the New Deal monetary measures, see Henry Mark Holzer, "How Americans Lost Their Right to Own Gold—and Became Criminals in the Process," *Brooklyn Law Review*, 39 (Winter 1973), 517–559.

must look at the events of the 1930s, not in their own terms, but interpreted as permanent amendments to the nation's monetary constitution, changes that help to define the monetary status quo of the 1970s.

Before 1933, the United States monetary system was an integral part of an international monetary order widely referred to as "the gold standard." In fact, however, neither the international nor the domestic monetary system adhered closely to the classical gold standard rules. Up until Great Britain's departure from the gold standard in 1931, a shaky semblance of international monetary order was maintained, but this was largely due to the fair weather cooperation of central banks superimposed on a single national system, that of Great Britain, which inspired worldwide trust and confidence. But a fatal deficiency in both the international and the domestic monetary systems was the fractional reserve base for monetary issue, a characteristic which made the whole edifice, national and international, dependent for its viability upon widespread confidence.

Nonetheless, there were features of the pre-1933 monetary constitution that appear extremely attractive, especially when set against criteria of individual freedom. The value of the dollar was defined by a specific gold content, $20.67 per fine ounce, a value that had not changed significantly since 1972, although inconvertible currencies circulated from time to time during the 19th century. The currency issue of the Federal Reserve System was limited to a defined multiple of the gold base. Currency was freely convertible into gold, both domestically and internationally. There were no restrictions on the ownership, purchase, and sale of gold in any form. The prevailing mythology of "sound money" exerted significant constraints on the behavior of individuals, in both their private-choice and their public-choice capacities. Individuals, firms, and governmental units could enter into voluntary contracts that provided for deferred payments of specified amounts of gold, or of currency equivalent to specified amounts of gold.[2]

All of this was swept away by a series of New Deal actions in 1933 and 1934.

We shall summarize these actions in Section III, but in order to discuss the monetary situation of the 1930s adequately, and especially to evaluate alternative remedial measures, the conceptual underpinnings of the "gold standard" must be summarized. This requires the application of some fundamental elements of monetary theory. We begin at a very basic level, with apologies to informed readers.

Consider first a country with no international trade in which only gold is used as money. For simplicity, assume that there is a fixed quantity of gold for monetary use; gold is neither being mined nor used for nonmonetary purposes. In this setting, as real income (and product) changes, due to productivity shifts or otherwise, the price level will tend to adjust in an opposing direction. If real product increases, more goods and services will be available, and in order to maintain employment and output, sellers will find it necessary to lower the prices of nonmonetary goods. With the fixed money stock, turnover of the increased supply of real goods at unchanged prices would require an increase in the velocity of money, or, to put the same thing differently, would require a decrease in the ratio of monetary assets to income. Behaviorally, individuals would seek to adjust away from such a disequilibrium, if indeed they should find themselves in such a situation. When the level of prices falls by roughly the same percentage as real

2. During the Greenback period, the U.S. Supreme Court had upheld the validity of specific gold clauses, even when these had the effect of distinguishing classes of monetary obligations. *Bronson v. Rhodes* (1869), 74 U.S. (7 Wall.), 229.

income has increased, the initial ratio of money (gold) to income would be restored.[3]

The same process also works for a decrease in real income or product, of course in the opposite direction. If a smaller quantity of real goods and services is available, and if prices do not change, people will find themselves with excessive holdings of gold. They will want to spend more money on goods because aggregate gold holdings will become excessive. This willingness will be implemented through a higher level of demand which will induce a rise in prices, an increase that will restore the desired ratio of gold to real income.

Let us now consider a country that uses only gold as money, but with an economy that is linked by international trade (conducted in gold) to the economies of other countries. The domestic adjustment mechanism described above will be available, but if the country holds a relatively small part of the world's gold, much of the adjustment will now occur through international gold flows. To a first approximation, one may say that so far as a single small country is concerned, the price of gold in terms of other commodities is determined in world trade markets. At the market price, a single country can have as much gold as it wants. If real income in one country rises, prices need not fall to maintain the ratio of money to income because additional gold can be imported to accomplish the same result. A slight decline in the price levels of all countries provides the additional gold to satisfy the demand of the one country. The maintenance of equilibrium in international trade requires that the adjustment take this form, because the price of gold relative to other com-

3. Prices and wages may, of course, be sticky and there may arise quantity and employment adjustments during the period between an initial change and a subsequent monetary equilibrium. It is precisely for this reason that monetary and macroeconomic theorists have objected to monetary systems which require that the price level fall secularly with increasing productivity in the economy.

modities must be the same in all countries. While the international linking of countries reduces the price level adjustment that a country must undergo in response to a change in its real income, it also forces any country whose currency is linked to gold to share in the price level response to income changes throughout the world.

Let us consider what happens when a country uses as money a combination of gold and paper currency that is backed by and freely convertible into gold. If people are confident that the paper currency will not lose its value, they will find it convenient to hold most of their money in currency, because it is easier to store and to transport than gold. But gold will be demanded by people engaged in international trade, by people who like the look and feel of gold, and by people who think that there is a nonnegligible probability that paper currency will lose some or all of its value. If the system is not disturbed by fluctuations in confidence, a gold-linked paper currency for a single country will be superior to a monetary system that relies exclusively on gold. Not only will the economies of storage and transport associated with paper be achieved, but also the country will be able to obtain the money for its system without having to buy gold from the rest of the world. (From the perspective of the world as a whole, the latter economy is illusory, for it merely reduces the value of the stock of gold.) To a first approximation, a single country's price level will be the same whether it uses only gold as money or a combination of gold and a gold-linked currency. Equilibrium in international trade will require that the prices of internationally traded items in the one country, expressed in gold, be the same as their prices in other countries. The only difference that the use of a paper currency makes is that it reduces the demand for gold, which is equivalent to saying that the world price level expressed in gold will be higher to the extent that countries supplement gold with paper currency.

If one makes the more realistic assumption that the quantity of gold is not fixed but rather is augmentable at increasing cost, there will be little difference in the response to sharp shifts in the demand for gold, because the stock of gold will be many times the annual production. However, in the long run the stock of gold will increase whenever there are deposits that can be mined at a cost below the monetary value of gold, which value will depend on the extent to which paper augments gold as money. The long-run equilibrium and the adjustment to it are susceptible to the traditional economic analysis of supply and demand with stocks and flows.

As a final elaboration of the monetary system, consider the consequences of supplementing paper currency that is partially backed by gold with demand deposits in banks that have fractional reserves of currency and gold. This institutional change will further reduce the ratio of gold to the total supply of money, which has both advantages and disadvantages.

The profit from substituting paper and demand deposits for gold will accrue to the entities (whether governments or private firms) that have exclusive permission to issue currency and demand deposits. The issuer receives valued goods or promises for mere pieces of paper it prints or for bookkeeping entries. The issuer may promise to reexchange the currency or deposits for gold, but unless it pays interest on its liabilities or keeps reserves equal to 100 per cent of its liabilities, the issuer is receiving an interest-free loan of indefinite duration, in an amount equal to that part of its currency or deposits that is not backed by reserves. If the issue of currency and demand deposits is not restricted, and issuers are allowed to compete, they will offer their customers enough extra services so that investment in this activity will yield the same return as in other activities.

The quantity of currency and deposits (as measured by

their gold value) is in any case limited by the demand for cash balances. If there is more currency and deposits than people wish to hold, they will try to spend their excess holdings. This will drive domestic prices up. People will begin to find foreign purchases more attractive than domestic ones, and will convert some currency and deposits to gold for international trade. To check the gold drain, issuers will have to reduce the amounts of currency and demand deposits, until domestic prices fall to a level that restores equilibrium in international trade. If the demand for gold continues to the point that an issuer's reserves are threatened with exhaustion, more drastic measures are required. One solution that has often been applied is a suspension of convertibility. The issuer says, in effect, "I'll give you gold when I can; in the meantime you'll just have to be satisfied with paper." Sometimes convertibility is reestablished when the bank (by selling assets) or the treasury (by raising taxes) acquires enough gold from the rest of the world; sometimes banks go bankrupt and treasuries simply renege by devaluing their currencies.

To guard against the need to suspend convertibility, a conservative government or private bank would try to keep a target level of reserves, often in some ratio to outstanding currency. If people demanded gold from its reserves, the bank would cut back its currency and deposit issue by some multiple of the size of the gold drainage. If people supplied the bank with additional gold, it would expand the currency and deposit issue, again by a multiple of the change in gold holdings. A bank would achieve the desired change in currency and deposit issue by changing its lending: To expand the issue, more loans would be made; to contract, fewer new loans would be made and old loans would be called or not renewed. Currency and deposit expansion would be associated with lower than average interest rates, and contraction with higher than average rates.

Beginning in about the 1870s, the quasi-public central bankers felt that it was not necessary for them to transmit every shock that they experienced to the domestic economy. They felt that they could successfully "lean against the wind" If an accumulation of reserves was seen as temporary, a central bank might refrain from expanding its lending. If a depletion of gold reserves was predicted to be short-lived, it might not contract loans outstanding. To the extent that bankers could be successful in such efforts, domestic interest rates would be somewhat more stable than under a regime in which the conservative gold-standard rules were strictly observed.

If international events provoke an outflow of gold that does not reflect a basic excessive supply of money, but rather a short-term fluctuation that will be reversed in a few months, then by permitting reserves to absorb all of the fluctuation rather than transmitting the fluctuation to the money supply through the banks, the monetary authorities can save the economy the costs of an unnecessary contraction and expansion. When bankers are required to contract their lending, businessmen must reduce their operations. Some firms will lay off workers. Other firms may face bankruptcy because they cannot raise funds to meet their obligations to creditors. The bunching of layoffs into a single period makes it all the more difficult for individual workers to find new jobs.

Furthermore, the reduction in the money supply associated with the credit contraction will generate a reduction in the equilibrium price level. Unemployed workers will typically find that they can obtain work only by accepting money wages lower than those that they expect to be able to secure. Some employed workers may find that they can hold their jobs only if they accept reduced wages. If workers do not perceive the possibility that the whole price level may be falling, so that lower money wages may retain the same real purchasing power as before, they may decide to keep looking for jobs that pay more acceptable wages. Similarly, people with products to sell may accumulate substantial inventories before offering the price discounts needed to move their merchandise.

An unanticipated expansion will have opposite effects. Employers will be unable to find workers as easily as usual and will raise wages. Inventories will be unusually depleted and shortages will develop until prices are raised. These processes depend critically on expectations about the rate of changes of prices. If people generally believe that prices are rising at 4 per cent per year, they may regard circumstances that limit the rise in the prices and wages that they receive to 2 per cent as just as unacceptable as a fall in wages and prices. If they perceive prices to be generally falling, they may find a corresponding fall in the prices and wages that they receive just as acceptable as stability in their own prices when prices are generally stable.

One more element of monetary theory is needed to complete the picture, and that is the role of money as a standard of deferred payment. When two people make a contract that involves payment of a definite amount of money at a time after the contract is made, there is some risk that the money will not be worth what they think it will when delivery occurs. The contract implicitly assigns this risk. The payer risks the possibility that money will be worth more at the time of payment than he had anticipated when agreeing to the contract, so that he would be paying more in real terms than he had anticipated. The payee risks the possibility that prices will rise unexpectedly, so that he would be receiving less than he anticipated. In a system in which gold is the only money there is a rough symmetry in the treatment of debtors and creditors; neither is protected against waves of hoarding and dishoarding and against long-term swings in gold production, but both are substantially protected against short-run disruptive and unpredictable changes in the supply of money. The supply of money is jointly determined by the behavior of all persons producing and using money, and not by any governmental authority.

When a gold reserve system is substituted for the idealized gold coin standard, and when a commercial banking system organized on a fractional reserve basis is superimposed on this, the relatively modest, symmetric uncertainties of a commodity market are replaced by the more complex uncertainties of confidence swings and political processes. There is first the political risk of a change in the gold value of the monetary unit. Excessive expansion of paper currency will put central bankers in a position where they must choose between the painful unemployment generated by monetary contraction and the embarrassment of a devaluation. Political expediency often lies with the latter. If the gold value of paper currency could somehow be made inflexible (or if creditors are permitted through gold clauses to insure against the possibility of devaluation), and if the central bank or treasury is required to maintain a stable ratio of paper currency to gold coin or bullion, then the individual creditor is substantially protected against significant inflation. But the potential debtor is not comparably protected against deflation, because wholesale conversion of currency into gold can result in a multiple contraction in the money supply as bankers seek to maintain the ratio of reserves to currency. This vulnerability of a gold reserve system is aggravated when a fractional reserve deposit banking system is superimposed on the currency issue authority. As individuals begin to doubt the soundness of the system and seek to exercise their rights of convertibility, they exchange bank deposits for currency (which serves as a reserve for bank deposits) and/or gold (which serves as a reserve both for currency and for bank deposits). Dramatic reductions in the effective supply of circulating media result in this way, due solely to changes in expectations.

If such a crisis of confidence affects only a single country that respresents just a small part of the world gold system, the consequence need not be prolonged monetary distress. The rest of the world will offer a relatively elastic supply of gold, imports of which will permit banks to pay gold to anyone who wants to hold it.[4] The only cost is an increase in other exports or a decrease in other imports. However, a world monetary system built in such a way on fractional reserves is vulnerable to a rapid succession of bank failures and currency devaluations in major countries. A combination of such confidence-shattering events makes it impossible to satisfy the world-wide public demand for currency and gold, given a low level of confidence in banks and national currencies, without a substantial contraction in income. The accompanying reduction in the price level drastically increases the real burden of deferred payments, to the financial ruin of debtors.

III. The New Deal Monetary Actions

With the above theory as background, the crisis that Roosevelt faced when he began his term in office in March 1933 can be more readily understood, and we can explore more fully the alternatives for action along with the measures actually taken.

Commencing in late 1930, the United States suffered through a series of banking crises. Many banks failed during

---

4. See Milton Friedman and Anna J. Schwartz, A Monetary History of the United States, 1867–1960 (Princeton University Press for the National Bureau of Economic Research, Paperback Ed. 1971), pp. 108–110, 158–162, on the panics of 1893 and 1907 for examples.

each crisis, and the Federal Reserve System did little or nothing toward restoring liquidity in the economy. The most severe of these crises took place in February 1933, just before Roosevelt's inauguration. On March 5, 1933, one day after assuming office, President Franklin D. Roosevelt, invoking the questionable authority of a 1917 Trading with the Enemy statute, declared a national bank holiday. All banks were closed, and they were prohibited from paying out gold or dealing in foreign exchange. This executive action was confirmed by Congress on March 9, 1933. Banks were allowed to reopen gradually, but the prohibition on dealings in gold and foreign exchange remained in force, apparently based on a fear that both an internal and external gold and foreign exchange drainage would accompany convertibility. One month after the bank holiday, the President issued an executive order that forbade the hoarding of gold and required all banks to deliver gold stocks to the central banks, at the parity of $20.67 per fine ounce. In June 1933, Congress passed a joint resolution abrogating gold clauses in all contracts, past and future, and in February 1935, the Supreme Court upheld this action by Congress, at least with respect to all private contracts

In January 1934, the Gold Reserve Act was passed, authorizing the President to reestablish a gold value of the dollar at a level between 50 per cent and 60 percent of its former value, and President Roosevelt immediately announced that the dollar was to be redefined by a new gold price of $35 per fine ounce, representing a reduction in the gold value of the dollar to 59 per cent of its former level. In effect, the United States returned to a gold related monetary system after this date, but the restrictions on private ownership and trade in gold were retained, along with the prohibitions on gold-clause contracts. Gold was bought at the new price only by governmental authorities, and, as a result of international agreement in 1936, sales restricted to cooperating foreign central banks were made. The devaluation insured that, for the succeeding thirty years, the domestic monetary structure would be divorced from the internal and external discipline imposed by the operation of a gold standard. Nominally, external gold movements could exert feedback effects on internal monetary policy, but these were not relevant for the United States until the 1960s.

In 1933, continued adherence to the international gold standard, as it was then operative, might have prevented or substantially delayed the internal monetary expansion that was so urgently required. The time was ripe for dramatic shifts, but in one sense, problems were created by the failure of political leaders to act boldly enough. A shift from a gold-based to a national fiat currency could have been accomplished readily. Convertibility could have been suspended, with respect to both gold and currency, and new currency could have been issued that was explicitly divorced from gold. Had these steps been taken, an independent monetary standard would have been born. In this case, there would have been no need to call in gold held outside the government, to prohibit ownership and trade in gold, or even to abrogate contracts made in gold. The market price of gold in the United States could have been allowed to find its own level, and sale of the government's stock of gold might have made the price relatively low, so that gold clauses might not have been burdensome.

Instead the government suspended convertibility while keeping gold out of everyone's reach. It is not clear whether this was done as a prelude to devaluation, or out of deference to a gold mythology that presumed a currency to be sound because it had gold backing, even if no one could have the gold.

So long as the link between the dollar and gold was to be retained there were several pragmatic political reasons for the accompanying restrictions. But these must be separated into two sets. The initial prohibitions on convertibility could have been justified by a belief that individuals and firms, both domestic and foreign, had lost confidence in the United States monetary structure and that they would have, if given the opportunity after March 1933, attempted to increase their holdings of gold. The suspension of convertibility was, however, accompanied by the prohibition on trade in gold among private persons and firms and on private contracting in gold, along with the abrogation of former contracts that contained gold clauses. This set of restrictions can most readily be rationalized on the basis of an intent to return to gold at a higher parity. And it is in this respect that the whole policy framework of the early New Deal seems most suspicious. By calling in all gold stocks in exchange for the dollar equivalent in currency at the old parity, the government put itself in a position to secure 100 per cent of the profits from any devaluation that it might introduce. To the extent that private persons held title to gold, devaluation would have provided them with windfall gains. In fact, through its policy of asserting title to all gold stocks, and by calling in all gold at the preexisting parity, the government made a profit of almost $3 billion on the dollar devaluation.[5]

Essentially the same rationale can be extended to the initial actions which abrogated the gold clauses in contracts that existed in 1933. The holder of a deferred claim expressed in gold was in a position equivalent to the holder of gold itself. And windfall gains would have been secured if the dollar price of gold had been increased. Windfall gains would have accrued to creditors holding contracts with gold clauses while windfall losses would have accrued to debtors in this

---

5. In the political temper of the times, this was probably more widely acceptable than a policy that would have allowed "speculative" profits to be made by private persons. Politicians were successful in shifting much of the blame for economic conditions away from government and onto private decision-makers.

set of agreements. The distributional consequences could have been prevented only by an abrogation of such clauses, insuring that contractual obligations would be met in units of current currency value.

With the intent of devaluation understood, the government's action in restricting gold ownership, trading, and contracting in 1933 can be explained. But a different question emerges after the January 1934 devaluation. Once the dollar had been redefined at its new and higher gold content, what was the basis for continuing the restrictions on voluntary contract? There seem to be three possible explanations. First of all, during the 1920s central bankers had come increasingly to the view that internal or domestic circulation of gold was not desirable. This position was based on a desire to maximize the monetary potential of gold by confining it to central banks and on an unwillingness to allow the existing gold standard to be opened up to the internal discipline and possible instability that individual ownership of gold would impose.[6] The shift toward gold bullion and gold exchange standards found ample support in the pre-1933 discussion. A second possible reason for the failure to remove restrictions on gold ownership and use following the devaluation was probably insecurity about the effects of devaluation itself. As the subsequent gold inflow amply demonstrated, there need have been no fear that the new value of gold was set too low, but Roosevelt Administration officials and advisers may not have predicted this effect. Hence, the restrictions may have been held on in the possible anticipation of still further devaluation for the same reasons as those that prompted the intial actions. Finally, and most important-ly, the restrictions on individuals' dealings in gold may have

---

6. Cf. Gustav Cassell, *The Downfall of the Gold Standard* (New York: Augustus M. Kelley, 1966), pp. 15–18. (First published in 1936).

been viewed as one means of preventing the multiple contractions in aggregate money stocks that had produced the banking crises. Even if nothing was to be done about fractional reserve banking, and hence the inherent instability involved in potential shifts between high-powered and low-powered money, there may have been the feeling that removing the prospects for conversion into very high-powered money would tend to generate confidence in the dollar currency itself. The administration decision-makers did not foresee the importance of Federal Deposit Insurance in reducing this type of instability in the total system.

In retrospect, it seems clear that full convertibility into gold could have been restored in 1934, at the $35 price. Domestic ownership and trade in gold could have been reintroduced without any of the fears coming true, at least for the years before World War II. The policy mix as it was implemented offered the worst of both worlds in a sense. Individuals could not avail themselves of the protection, the predictability about monetary matters that the traditional gold standard had possessed, or at least had been thought to possess. At the same time, they were not allowed to make ordinary private contracts in gold, as would have been possible under a truly independent national fiduciary standard.

The monetary constitution, as it emerged from the New Deal, was a makeshift affair, consistent with no single conceptual framework. The international adjustment mechanism, tied as it was to gold as a special commodity with a defined dollar price, seemed to justify the prohibition of possession, purchase, and sale of gold by Americans. As a result, changes that might have been temporary aberrations from the country's long-standing traditions of free voluntary contract and free markets were cemented into the legal order, bringing with them the precedent-setting potential for still other interventions with freedom of contract.

IV. The New Deal and Constitutional Law

We have deliberately referred to the New Deal monetary changes as "constitutional," and we have used this word in its generalized meaning. A "constitutional" change is one that modifies the "rules of the game," the institutional framework within which both public and private decisions are made and actions taken. A "constitutional" change is intended to be, and is understood to be, quasi-permanent; it is expected to remain in force over some indefinitely long time period. In this most general usage of "constitutional," there is no explicit reference to constitutionality in a specific legal or historical setting. Nonetheless, the generalized conception must offer the principles upon which constitutional law is normally distinguished from other branches of law.

We should have expected that monetary changes that were understood to be "constitutional," in the generalized sense of the term, would have aroused controversy in constitutional law per se in the United States. This controversy was concentrated in a series of cases that challenged the constitutionality of the abrogation of the gold clauses in contracts. The Supreme Court considered these cases as a group, and explicitly addressed itself to the legality of the Joint Resolution passed by Congress on June 5, 1933, the resolution that abrogated gold clauses in all contracts, public and private, past and future. In February 1935, the Supreme Court, in a set of five-to-four decisions, upheld the legality of the gold-clause abrogations with respect to all private and local government obligations, and hedged on its judgment with respect to the obligation of the federal government itself.[7]

---

7. See *Norman v. Baltimore and Ohio Railroad Co.*, 294 U.S. 240; *Nortz v. United States*, 294 U.S. 317; *Perry v. United States*, 294 U.S. 330.

In this section we shall examine the judgment of the Court, both with reference to criteria of monetary and economic theory and with reference to the conception of contract.[8]

Initially, it is necessary to stress that money is different in its legal characteristics from other aspects of "law" even when ideally considered. The written Constitution of the United States gives to the Congress the power to regulate the value of money, and constitutional law applicable in other countries has been interpreted as giving to national governments comparable powers to those that were explicitly delegated to the United States federal government. In the United States and elsewhere this has been interpreted to mean that the legislative branch of government has the authority, as constituted, to make genuine changes in the "constitution" with respect to money.[9] Legislatures can, effectively, change the fundamental "law" in this area, something that they cannot do in the more general setting of rulemaking, or at least, they were not considered to be able to do in the period of the New Deal.[10] (One problem in examining the whole

8. These events stimulated a flurry of discussion among constitutional lawyers, and the law journals and reviews published numerous papers in the months immediately preceding and following these decisions. A selected listing of these papers is as follows: John P. Dawson, "The Gold Clause Decisions," *Michigan Law Review*, 33 (March 1935), 647–683; John Dickenson, "The Gold Decisions," *University of Pennsylvania Law Review*, 83 (April 1935), 715–25; Henry M. Hart, Jr., "The Gold Clause in United States Bonds," *Harvard Law Review*, 48 (May 1935), 1057–99; Arthur Nussbaum, "Comparative and International Aspects of American Gold Clause Abrogation," *Yale Law Journal*, 44 (November 1934); J. Roland Pennock, "The Private Bond Case as a Postponement of the Real Issue," *University of Pennsylvania Law Review*, 84 (December 1935), 194–211; Russell Z. Post and Charles H. Willard, "The Power of Congress to Nullify Gold Clauses," *Harvard Law Review*, 46 (June 1933), 1225–57.

9. Persons with a more conservative view have maintained that the only power delegated to Congress was the power to specify a meaning of the dollar in terms of gold or some other commodity, but this view has not prevailed. See John Sparks, "Notes on the Legal Aspects of Gold Ownership" (Mimeographed), Hillsdale College, Hillsdale, Michigan, 1974.

10. As Gerald T. Dunne notes, the absence of more specific definition of

Continued on p. 30

set of policy measures and the court decisions from the vantage point of 1975 is that of overcoming the profound differences in constitutional attitude that have taken place over the forty-year period.) What this means is that "amendments" to the effective monetary constitution can be made by ordinary legislative majorities of national governments. This power to modify the real constitution was explicitly set out in the written legal Constitution in the United States, and comparably interpreted in other nations.[11]

This explicit delegation of power to the Congress, and through Congress to the Executive, removed from the judiciary any seriously considered legal issue as to the legitimacy of most of the New Deal monetary measures. The prohibitions on the ownership of gold were subjected to only minor legal challenges because gold was acknowledged to be the basic monetary commodity.[12] Hence, any policy directly relating to gold was held to be within the explicit delegation of power to regulate the value of money. The federal government would have been successfully challenged in 1933 had it prohibited the private ownership of any nonmonetary metallic commodity, say copper or zinc. (Once again, we must concentrate on the legal situation of the early 1930s. After

the monetary authority in the written Constitution of the United States has been one element that has allowed governmental authority to be expanded, perhaps beyond all limits foreseen by the Founding Fathers. See Gerald T. Dunne, *Monetary Decisions of the Supreme Court* (New Brunswick: Rutgers University Press, 1960), p. 1.

11. The Principle of "nominalism" has been adopted as the basis for the "law of money" in most Western nations. This principle states that a monetary unit (a dollar or pound) is what the national government declares it to be. Hence, monetary obligations are fulfilled by payment in monetary units, as defined at the time of repayment. For a detailed discussion, see F. A. Mann, *The Legal Aspect of Money*, 3d ed. (Oxford: Clarendon Press, 1971), pp. 76–96.

12. For a discussion of the unsuccessful challenges, along with citations, see Henry Mark Holzer, "How Americans Lost Their Right to Own Gold—and Became Criminals in the Process," *Brooklyn Law Review*, 39 (Winter 1973), 517–559. These challenges were not accepted for review by the Supreme Court.

the personnel and the attitude of the Court changed in the direction of the legal position represented by Justice Felix Frankfurter, the Supreme Court became much more reluctant to find any legislative action to be outside constitutional powers.)

Those groups and interests who opposed the New Deal monetary measures were limited, in their potential constitutional challenges of import, to the abrogation of the gold clauses in contracts. These challenges were made with the explicit acknowledgment that the Congress had the power to regulate the value of money. The challenges were reduced to the subsidiary question of whether or not the abrogation of the gold clauses was necessary to the acknowledged power to regulate the value of money. In its ruling, the Court held that this action was a necessary part of the larger set of actions in the exercise of this acknowledged legal power of the Congress. It is this basis of the Court's decision that we must evaluate. In so doing, we must apply both modern monetary theory and monetary theory as this was understood by the Court at the time of its decision.

We must begin by an examination of the purpose of the monetary measures of the New Deal, considered as a combined package. Why was it necessary to take drastic action in 1933? The banking structure was in chaos; confidence in the security of bank deposits had nearly disappeared; and individuals were seeking to convert deposit claims into currency. To the extent that *some* individuals were successful in this, a multiple contraction in the nation's stock of money was required. But without additional high-powered money, *all* persons could not accomplish this, because of the fractional reserve basis for the commercial banking system. The attempt by many persons to liquidate deposits was, literally, producing insolvency for the whole financial system. The epidemic of bank failures might have undermined confidence in currency itself, producing attempted conversion of currency into gold.

The first requirement of any policy was, therefore, to stop the attempted shifts into high-powered money and to restore confidence in the system, as it then existed. The suspension of convertibility was a reasonable step, and this clearly seemed to lie within the authority of the government.

Having suspended convertibility, however, did it follow that all trade in gold should be prohibited, and that subsequently, all gold be called in and private ownership forbidden? It might be plausibly argued that, so long as gold was allowed privately to circulate, individuals would not reattain confidence in using currency or bank deposits, that confidence in the system would be restored only if the private circulation of gold, independently of the convertibility privilege, could have been prevented. For purposes of this analysis, let us accept this argument as one that was empirically justifiable.

This would have provided the basis for the calling in of gold held privately, and the prohibition of private ownership.[13] In this action there need have been no confiscation of value, however, and no redistribution of wealth from one class to another. It is only when we move to the next step beyond this that such questions arise. The economy desperately needed an increase in aggregate demand, in purchasing power. Deflation had occurred; prices were much lower than 1929 levels. Inflation was positively, indeed urgently, needed, at least to the extent of reattaining 1929 levels of incomes and prices. Monetary means of promoting this inflation required that the aggregate supply of money be expanded.[14] The Thomas Amendment to the Agricultural

13. There was an existing legal justification for these actions. In its 1869 decision upholding the constitutionality of a tax on state bank notes, the Supreme Court stated that Congress could restrain the circulation as money of anything that it had not specifically authorized. *Veazie Bank v. Fenno*, 8 Wall 533 (1869).

14. In retrospect, the need for expansion of the money supply seems clear, although at the time there was also a fear of excessive inflation, as in Germany in the early 1920s, so that measures to expand the money supply encountered considerable opposition.

Adjustment Act, passed in 1933, authorized the issue of new currency, but the idea of devaluing the dollar in terms of gold was widely interpreted as the most effective means of securing an expanded monetary base for inflation. Only a month after the suspension of convertibility there were indications of an intent to raise the dollar price of gold. And, of course, suspension of convertibility itself created speculation concerning future devaluation. It is at precisely this point that profound issues of equity as well as contract were introduced. If the dollar price of gold was to be arbitrarily increased, and if private holders of gold were to be required to turn in all holdings to the Treasury, what price did they "deserve" to get for their stocks: The old parity of $20.67, or the new parity of $35 per fine ounce? In the first case, the private holders of gold would have suffered an opportunity loss in the amount of more than $14 dollars per ounce, although they need not have suffered historical losses in any accounting sense. The government would have gained all of the profits on the revaluation of gold, and not only the profits on the revaluation of its own holdings. In the second case, all persons and units, including the Treasury, who happened to hold gold stocks at the moment of the depreciation would have secured windfall gains. We know that the first of these two alternatives was chosen; by calling in all gold and paying for this at the old parity, the government tried to secure *all* of the profits from devaluation of the dollar.

We shall return to these alternatives, but let us now look at the relevance of the gold clauses in outstanding contracts at the time of the change in the gold price. The private owner of an ounce of gold was required to turn this in to the Treasury at the old parity. He gained nothing from the depreciation. As a matter of simple equity, it might have seemed that the holder of a deferred claim to gold, a creditor holding an obligation containing a gold clause, should not have been differentially favored over the holder of the monetary commodity itself. To call in gold at the old parity while, at the same time, continuing to honor the letter of contract through the gold clauses might have seemed contrary to justice and to law. Some such attitude might have informed the thinking of the Roosevelt Administration when the Joint Resolution was passed and also that of the majority of the Supreme Court when this resolution was upheld.

The critical questions should have been raised, not on the gold clause contracts in isolation, but on the distribution of the gains or profits from dollar depreciation. The holder of gold and the holder of deferred claims to gold might have been treated equitably by paying *both* at the new parity, that is by paying holders at $35 per ounce and honoring all gold clauses in obligations outstanding. The dissenting minority of the Court held a weak reed when they limited themselves to the gold-clause abrogations while acknowledging the authority of the government to pay holders of gold coin or bullion at the old parity. Had the minority stood willing to bring the latter into dispute and to question the government's power in this respect, they could have constructed a more convincing argument.

Let us return to the alternatives noted above. In one sequence, holders of gold coin and gold bullion are paid at the newer and higher price, set at the time of the devaluation. All gold clauses in contracts are honored. In the other, and historically descriptive sequence, holders of gold coin and gold bullion are paid at the old, and lower, price, and all gold clauses in outstanding contracts are abrogated. The question that we must ask is whether or not this sequence was necessary if the government, in the situation confronting it in 1933, was to fulfill its constituted authority to regulate the value of money. If we answer this specific question affirmatively, we must conclude that the majority opinion of the Court was correct, despite its apparent negation of long-standing principles relevant to the sanctity of contract. If we answer this specific question negatively, we must conclude that the majority of the Court erred in its judgment on its own grounds and that serious and wholly unnecessary harm was wrought to long-standing constitutional principle, either due to a misunderstanding of monetary theory or to a concealed attempt to achieve distributional objectives that could not have been embodied in the monetary clause of the Constitution.

What did the Roosevelt Administration, and its rubber-stamp Congress, seek when it went beyond the suspension of convertibility to devaluation? As we have already noted, one plausible objective would have been that of generating a domestic inflation in prices and incomes.[15] As suggested above, this would have required that the supply of circulating medium, currency and bank deposits, be increased. We must, therefore, examine the potential effects of devaluation in this respect. Our question becomes: Would the alternative sequence have prevented this increase in the supply of circulating medium and, hence, have interfered with the accomplishment of this basic governmental purpose? Analyzed in this way, in the context of the sequence historically followed, we reach an interesting, and somewhat surprising, conclusion. As we shall try to demonstrate, the objective sought by devaluation could have been better achieved by the alternative than by the actual scenario. That is, domestic prices and incomes might have been boosted more promptly and more effectively if holders of gold coin and bullion had been paid at the new and higher parity and if all gold clauses had been honored.

15. A more traditional reason for currency devaluation has been deterioration in a nation's gold and foreign exchange reserve position. But the United States devaluation was almost unique in this respect; there was no danger to the nation's foreign reserve position prior to the devaluation. On this, see M. Palyi, *The Twilight of Gold, 1914–1936* (Chicago: Henry Regnery, 1972), p. 280.

Consider the following possible chain of events. Suppose that, in 1933, the government should have suspended convertibility, and, simultaneously, raised the price of gold to $35 per fine ounce, while calling in all gold coin and bullion at the higher price and outlawing subsequent private ownership. In order to purchase gold, the government would have drawn from its deposit balances in the Federal Reserve Banks; checks drawn on these balances would have been transferred to private parties in exchange for gold. To cover these withdrawals, the government would have issued gold certificates on the newly-acquired stocks and these certificates would have been transferred to the Federal Reserve Banks in exchange for newly-created government deposits, just sufficient to replace those used up in the initial gold purchasers. In the gold-purchase transaction, as such, the Treasury would neither have gained nor lost. As private parties deposited these government checks in their own commercial banks, these banks would clear these through their Federal Reserve Banks. And, since these government checks were drawn against a Federal Reserve account, commercial banks would find that their own reserves had been increased, while deposits had increased by only some fraction of this amount. This would have provided the basis for a multiple expansion in the quantity of bank deposits.

Under ordinary circumstances this would, in itself, have been sufficient to generate inflation. But banks already possessed excess legal reserves, and although some pressure toward expansion in deposits would have been generated by an increase in reserves, there is no guarantee that the overall effect would have been sufficiently great. But there is a second and more direct force that would have been at work to spark an inflation in domestic prices and incomes. Persons and firms who had profited from holding gold at the time of devaluation would have found themselves holding more assets in the form of currency and bank deposits ($35 for every $20.67 held earlier) than they desired to hold. They would have attempted to shift out of monetary assets into nonmonetary assets. As a result, prices of nonmonetary goods and services would have increased.

To the extent that these two forces operated fully, and that the government maintaining the predevaluation ratio of money to gold, there would have been an increase in money income roughly proportional to the devaluation. In this setting, the holders of gold-clause obligations would not have gained relative to debtors. Since incomes would have been increased generally throughout the economy, the burden of payment of a gold-clause obligation would have been the same as before the devaluation and subsequent inflation. To the extent that the inflation in incomes and prices was not proportionately so large as the devaluation, debtors who had to meet gold-clause obligations would have suffered in real terms, whereas creditors would have gained. In this case, creditors would have been in the same position as those who held gold coin or bullion at the time of the devaluation.

We are not defending this alternative scenario in the sense that it is advanced as an "ideal" policy package for the setting of 1933. It seems unlikely that incomes and prices would have increased proportionately with the devaluation; supplementary measures toward expanding the supply of circulating medium would surely have been required. But we have constructed this alternative scenario in some detail here for comparative purposes. This scenario, had it been followed, would have generated *more* of the desired effects than that sequence of policy measures that were enacted and put into force by the Roosevelt Administration. Let us now examine the latter in more detail.

Convertibility was suspended; dealings in gold were prohibited; individuals were required to turn in all gold to the Treasury; private ownership was forbidden. But the price for gold coin and bullion was set at the old rather than the new parity. The government tried to guarantee to itself all of the profits from the devaluation. Distributionally, this might have been questioned, but this distributional difference, in itself, need not have prevented the attainment of the desired inflation in incomes and prices had not the government effectively sterilized the gold purchases. For gold purchased from private persons, the Treasury issued gold certificates and transferred these to the Federal Reserve Banks in exchange for federal deposit accounts. These accounts were drawn down to pay for the gold, as in the alternative scenario, but a profit was left over. If the government had utilized these profits to expand its rate of spending in the economy, the desired increases in incomes and prices might have taken place. Instead of this, however, the Treasury set aside $2 billion of the $2.8 billion profits in a special exchange stabilization fund, while $645 million was used to replace national bank notes. Of the total of $2.8 billion, less than $200 million was made available for general fund outlays by the federal government.[16] For the bulk of the paper profits, the disposition made effectively insured that this potential purchasing power could not be returned to the domestic economy and could not, therefore, become reserves of the banking system. The forces that would have been at work under the alternative sequence were, therefore, largely nullified. Private parties who previously held gold found themselves with the same share of their assets in monetary form and hence had no incentive to expand spending rates. And the government did not use the profits to expand its own spending rate. In effect, the devaluation effort per se was "wasted" almost completely; the opportunity to increase the money supply in 1933 directly through devaluation was missed, although the quantity of money did begin to increase rapidly in 1934.

16. Cf. Friedman and Schwartz, p. 471 n.

Events moved swiftly in the dramatic years of the 1930s. By the time that the Supreme Court considered the arguments in the gold-clause cases, during the October 1934 term, it probably was clear that the restoration of predepression incomes and prices was not imminent. As an inflationary measure, devaluation seemed to have failed. In this setting, and quite apart from the equity as between holders of gold and deferred payment obligations, the enforcement of gold clauses in contracts would have seriously damaged debtor interests. Estimates as to the quantity of gold-clause obligations outstanding ranged up to $100 billion. Debtors under these contracts would have been required to increase payments in real terms by some 69 per cent over what they might have legitimately expected to pay in the absence of the devaluation. The real burden of debt would have been increased rather than decreased by the enforcement of these clauses. Recognition of this was surely one influence on the Court's majority opinion.

In a sense, the revalued price of gold was wholly abitrary in the setting that actually occurred. Since the potential expansionary effects of revaluation, which would have increased incomes throughout the system and reduced the burden of debt designated in nominal dollars, were largely nullified by the actions of the Treasury and the Federal Reserve Banks, the arbitrary price set on gold was divorced from the domestic economy. In subsequent years, this price did generate an inflow of gold which was then allowed to provide a basis for gradual monetary expansion, but this was not the setting within which the Court's decision was made.

How do we answer the question posed above with respect to the Court's findings? Were the actions taken in 1933-34 necessary if the government was to fulfill its constituted authority to regulate the value of money? The analysis suggests that abrogation of gold clauses in contractual obligations was not a necessary complement to the authority of government to regulate the value of money in the setting of 1933, *provided* that the government itself should have acted consistently in furtherance of its announced objectives. But, as we have indicated, the government was far from consistent in its set of policy actions, and it effectively nullified the potential results of devaluation. In this situation, what could the Court have decided? The Court could not, itself, instruct the government in monetary theory, nor could it force the government to take the positive action that might have been required to accomplish its declared objectives. (Again, it is necessary to emphasize the difference in the presumed power of the Supreme Court in the 1930s and in the post-Warren years when the Court commenced to take initiatory action in the absence of legislative action, as, for example, in the apportionment decisions.)

There was one important element in the situation confronting the Court that we have not explicitly discussed, although it is, of course, related to the objective of the New Deal measures. In reading the opinion of the majority of the Court, it seems clear that the gold-clause abrogations would not have been upheld in a situation where the government devalued the currency from a stable basis with the purpose of securing profits for itself. As the Court stated, the purpose of the gold-clause obligations was to protect the creditor against declines in the real value of the obligation, that is, against monetary inflation subsequent to the initiation of the contract. But few, if any, holders of gold-clause obligations would have suffered reductions in real value through abrogation. The domestic economy had undergone significant deflation in incomes and prices in the crisis years after 1929. Inflation was required, and desired, in order to get back to a pre-Depression *status quo ante*. In advancing their claims for increased real value under the gold clauses, creditors were forced to argue that they should gain distributionally as a result of fortuitous economic circumstances, that they should be immune from general economic loss because of the particular structure in which contracts were written, and which was introduced for quite different reasons. Interpreted as protective devices against reductions in real value, the gold clauses were, in a sense, redundant in the circumstances of the early 1930s. Real values of all monetary obligations had increased substantially; enforcement of the gold clauses after devaluation would have added yet another increment to the real value increase accruing to holders of such claims. This element strengthened the judgment of the Court's majority. Perhaps more importantly for our purposes, the Court's decision cannot be read as legitimizing subsequent prohibitions on index contracts, whether in gold or other commodities or in some composite, that are designed solely to protect citizens against inflation from a stable or even increasing price level.[17]

V. Contractual Origins of Monetary Rules

Article I, Section 8 of the United States Constitution gives the Congress the power to "coin money, regulate the Value thereof, and of foreign coin, and fix the Standard of Weights and Measures." As we have noted, the problem faced by the Supreme Court in the gold-clause cases, as well as in earlier judgments, was one of determining whether the actions taken by Congress fell within this designated monetary authority. For the gold-clause legislation in particular, this included the determination as to whether specific contractual

17. B. Nowlin Keener suggests that this emphasis in the Court's majority opinion, written by Chief Justice Charles Evans Hughes, possibly reflects Hughes' sympathetic awareness of the ideas of Irving Fisher. In support, Keener mentions the interesting fact that in the early 1920s, Hughes had been an honorary vice-president of Fisher's Stable Money League. See *Transcript of Keener vs. Congress with Explanations* (Privately circulated, 1973), p. 22.

agreements (the gold-clause obligations) interfered with Congress's exercise of this stipulated monetary authority. If congressional action was explicitly taken within this authority, and if the Court upheld the Congress's factual assertion to this effect or made this factual finding on its own, the legal issue, strictly speaking, was settled. There was no need for the Court to go behind the actual statement in the written Constitution, no need for the Court to examine the appropriateness of the stipulated monetary authority, to analyze the possible meaning of the constitutional statement itself.

In this section, we shall attempt to look critically at the monetary component in the political constitution with the objective of deriving a delegation of monetary authority to legislative bodies from the idealized constitutional-choice calculus of persons in a constitutional convention. Once we have taken this step, we can then return to the language of the stipulated power in the United States Constitution, as written, and to the interpretations that have been placed on this power.

Consider a hypothetical setting for constitution-making. A group of persons is assembled for the purpose of agreeing on rules under which subsequent private and public behavior can take place. To the extent that individual economic positions in subsequent periods are predictable in advance, there would be little prospect of securing genuine agreement on rules, on "a constitution." In order to construct a setting within which genuine constitutional attitudes may inform individual behavior, we can take either one of two possible alternatives. First, we can postulate that the rules to be chosen are to apply over a sufficiently long time sequence to insure that individual positions in specific subperiods remain uncertain.[18] Second, we can postulate that the individual adopts

18. This is the choice setting postulated by Buchanan and Tullock in their analysis of constitutional decisions. See James M. Buchanan and Gordon Tullock, *The Calculus of Consent* (Ann Arbor: University of Michigan Press, 1962).

a normative attitude in which he places himself behind a "veil of ignorance," where he "thinks" himself equally probable of occupying any position in the society.[19] In both of these logical idealizations of a choice setting, the persons will be led to evaluate alternative rules and/or institutions on the basis of their general applicability and working properties, rather than in terms of the potential furtherance of any identifiable self-interest. "Fairness" and/or "efficiency" become plausibly acceptable criteria for judgment here, judgment which is applied to the procedure or process rather than to the particular outcomes generated.

We shall not elaborate on an analysis of the constitutional choice setting, since adequate treatment of this can be found elsewhere. What we want to do here is to apply this conceptualized contractual decision-making to monetary rules and institutions specifically. We want to ask: What sort of a "monetary constitution" might be predicted to emerge from a genuine social contract, hypothesized in the manner that we have indicated? Not knowing whether he will be rich or poor, creditor or debtor, banker or butcher, gold miner or wheat farmer, economist or lawyer, how would the individual, at this true constitutional level, evaluate alternative monetary arrangements that might be suggested or proposed?

To our knowledge, this question has not often been specifically addressed.[20] Beyond an initial definition of rights and the establishment of instruments to enforce these rights,

19. This is essentially the choice setting used by John Rawls. See his, *A Theory of Justice* (Cambridge: Harvard University Press, 1971).
20. In L. B. Yeager (Ed.), *In Search of a Monetary Constitution* (Cambridge: Harvard University Press, 1962), the authors of the separate essays were explicitly requested to analyze and present alternative monetary frameworks or constitutions. However, there was little specific treatment of the consistency of selected arrangements with an initial contractual settlement. Buchanan's essay provides a partial exception here; his discussion was implicitly based on the contractual model described above.

the "fair" or "efficient" constitution would include some specification of the rules for reaching collective or governmental decisions, rules that would also include a definition of the range over which governmental action might operate. With respect to monetary institutions, we must first examine the objectives to be served. What would be desired from a set of monetary rules, a monetary framework?

We may answer this question in terms of two separate functional objectives. There are genuine "public good" characteristics of money; agreement among persons concerning the thing to be used as money will greatly facilitate economic exchange. The delegation of the authority to *define* the monetary unit to the government seems to be a plausible outcome of a constitutional contract. The definition of what is to be treated as money in the society does not seem much different from the definition of what shall be the ordinary standards of measure, and it is no surprise that the power to fix weights and measures is found in the same sentence of the United States Constitution that provides for regulating the value of money. The definition of the monetary unit does not, however, directly imply anything about the value of money through time. A second functional objective that should plausibly emerge from constitutional contract is *predictability* in the value of money, or reciprocally, in the absolute level of prices. It is difficult to see how anyone should object to this attribute as a desirable feature of monetary order, and, if government is to be granted any authority beyond the definitional, regulation with a view toward insuring predictability seems the minimal attribute upon which all persons should agree. To the extent that persons and organizations, in their ordinary capacities as economic decision-makers, can predict the value of money, they need not introduce this extraneous element of uncertainty into their decision calculus.

We note that predictability, as an objective, does not

prejudice the constitutional process toward going further and selecting monetary arrangements that will produce either *stability* or *change* in the level of absolute prices through time. There remains the prospect of disagreement among those who mutually accept predictability as an objective. Some may seek a secular decline in prices; others, a secular rise; still others, stability through time. Potential differences here suggest that predictability defines the limits of consensual agreement at any genuine constitutional stage of choice. It is quite difficult to imagine why any persons would prefer uncertainty to predictability here.

If monetary arrangements are to be agreed on at the constitutional stage,[21] if monetary rules are to be among the set of quasi-permanent institutions of social order, the governmental power "to coin money," which means that of selecting and defining the monetary unit, and to "regulate the Value thereof" seems conceptually appropriate, *provided* that the regulation is understood to have as its objective the insurance of predictability in value. That is to say, the second part of this should be modified to read "regulate the Value thereof so as to insure predictability in this Value." This appears to be a meaningful derivation from genuine social contract. It would be difficult to imagine a constitutional statament that was intended to provide legislatures with powers to regulate the value of money for open-ended purposes, which might include the self-interest of the members.

We emphasize that these are minimal constitutional limits. Individuals may disagree sharply as to the instrumental means

21. Libertarian anarchists might argue that no role for government with respect to monetary arrangements need emerge from a true constitutional process. Implicitly, they are assuming that individuals would agree on a money commodity and that this spontaneous system would do more toward insuring predictability than one that provides to government any powers at all over money.

through which predictability might best be furthered. Even
if there is general agreement on the working properties of
alternative structures, disagreement may remain as between
the two major sets, those of automatic and those of managed
monetary systems. As we have defined the contractual or
constitutional setting, it is not possible for us to discriminate
between these two broad alternatives for monetary order.
Either one or the other, in any one of a large set of subvariants,
might emerge from a deliberative constitutional assembly,
or, as in the United States, the constitutional specification
may not include more than a broad statement of governmental powers.

Let us now examine the United States monetary and legal
history with these two plausible contractually legitimate objectives of a monetary framework in mind. We find that
there has, indeed, been ample recognition of the first, or
definitional, purpose. The general benefits of having the
central government define the monetary unit have been
recognized in much of the historical and constitutional
development. By contrast, we find little or no recognition,
or even much discussion, of the second contractually derivable
purpose, that of predictability in the value of money. As
noted, the written Constitution leaves this authority unspecified, and we can only speculate on the reason why the
Founding Fathers agreed to include the phrase "regulate
the Value thereof." In practice, before 1933 (and after) the
government has done little that was identifiable as falling
within this regulatory power, as interpreted here. In the
exertion of its power to define the monetary unit, in the
exercise of the delegated definitional authority, the government has, of course, exerted spillover effects on the value
of money. The silver question, which generated much political
controversy during the closing decades of the nineteenth
century, was, on the surface, one of using the government's
authority to define the monetary unit. Should gold be the

only money commodity? The settlement of this question was known, however, to have a major impact on the value of money, on the absolute level of prices in the domestic economy. The silver interests wanted inflation in the monetary base; the gold interests did not. Nonetheless, because the issues were discussed largely in definitional terms, there seems to have been little or no attention paid to the objective of regulation of monetary value.

On balance, it seems fair to assert that the United States government, in the exercise of its duly constituted authority, might be judged to have defaulted in its contractually-justifiable purpose of regulating the value of money with a view toward insuring predictability and hence toward enhancing the overall efficiency of social order.[22] We can make this judgment without at this point choosing sides between the managed and the automatic monetary systems, or among the many subvariants within each category, not to mention the mixtures. We need look only at the historical record to ascertain that the value of money has fluctuated widely and unpredictably through our history.[23] Regardless of the descriptive characteristics of the set of monetary rules and institutions that were in existence, the regulatory objective was not met. The changes in the basic monetary constitution that were implemented in the 1930s must, therefore, be evaluated against historical failure rather than success, failure of the government to meet its constitutional authority to regulate money's value, if this authority is interpreted or defined in a way that might make the delegation of regulatory power contractually legitimate.

22. Congress's failure to carry out its specifically designated role of regulating the value of money was the basis of B. Nowlin Keener's legal action. His suit was dismissed in federal district court and his appeal denied on the grounds that the court lacked jurisdiction. See, *Transcript of Keener vs. Congress with Explanations* (Privately circulated, 1973).

23. See Chart 62, p. 679 in Friedman and Schwartz, for a convenient summary of the record.

VI. Managed versus Automatic Monetary
Systems

The constitutional authority to regulate the value of money is sufficiently unspecified in the United States Constitution to allow the Congress to adopt almost any organizational structure. The Constitution does not discriminate between a managed and an automatic monetary system, and, as we have suggested in the preceding section, either of these basic forms might emerge from a genuine constitutional contract. This failure of precision may have served, in part at least, to generate the failures noted above. The monetary system of the United States, both before 1933 and after, has been neither a wholly managed nor a wholly automatic system.[24] It has represented a peculiar mixture of both forms, producing what may have been the worst features of both.

It is helpful to an understanding of the constitutional events of the 1930s to engage in some imaginary history. Let us suppose that the Constitution delegated to Congress the power "to issue fiduciary currency in sufficient quantity to insure predictability in the value thereof." This would have amounted to an explicit direction to organize and to operate a system of national fiduciary money, without relationship to any commodity, such as gold, and without implications as to the relationships between the national system and those of other nations. Let us suppose that this "pure greenback" system, with a fractional reserve banking component

---

24. Since 1971, when gold sales and purchases by the Treasury were abandoned, and the dollar allowed to float vis-a-vis foreign currencies, the United States comes closest to being "pure." With these changes, it is perhaps fair to say that, finally, the system has been transformed into a pure managed system.

appended,[25] had performed in roughly the same fashion over time as that system that we did have. That is to say, assume that this pure fiat system has failed to prevent dramatic deflation in incomes and prices in the Great Depression.

In such a setting, any desired increase in the amount of circulating medium could have been produced by straightforward currency issue at relatively little cost. There would have been no constitutional barriers in the way, and individuals could have claimed no violations of legitimate expectations in such governmental actions. Suppose, however, that some persons had made specific debt contracts which contained protective clauses that required the repayment of principal in terms of the value of a designated commodity or an index of commodity prices. These contracts, if existent, should have remained fully valid, since they would have borne no direct relationship to the governmental action in increasing the quantity of fiat currency. Under a system of fiat issue, therefore, there should have arisen no question about the abrogation of private contracts in furtherance of the government's power to regulate the value of money. Freedom of private contract could have remained inviolate, at least in this respect, and the Court need not have been placed in the position of choosing between competing basic principles of constitutional law.

Let us now change our imaginary history and suppose that the written Constitution delegated to Congress the power

---

25. Strictly speaking, there would be no basis for allowing fractional reserve banking in such a model. Since the issue of paper currency is relatively costless, there would be no need to economize on scarce monetary resources through fractional holdings of high-powered money. If, however, the fiduciary issue is limited, fractional reserve banking may emerge as an institutional means of exploiting the "publicness" efficiencies inherent in high-powered money. In a pure fiat system, therefore, fractional reserve banking may be tolerated, not because it is intrinsically efficient, but because its prohibition may prove to be costly.

"to define a monetary commodity and standard units thereof," with no further specification as to the regulation of value. This would have instructed the Congress to select a commodity, say gold, and to set a price for this in terms of dollars. This would, in turn, have implied free coinage, that is, a willingness of the government to mint and to certify units of the commodity.

Why would anyone, at a constitutional stage of deliberation, prefer a commodity to a fiat standard, an automatic to a managed monetary system? We need to go back to the predictability objective emphasized above, which we identified as one of the two objectives that might plausibly emerge from genuine contractual agreement among persons. The automatic or commodity system, whether the commodity be gold or something else, could only be preferred on the basis of some prediction that this would embody greater certainty about the value of money than would a fiat or managed system. Reasonable men may disagree here, and those who opt for a commodity standard do so because they predict that governmental decision-makers will not, intentionally or inadvertently, hold to the guidelines that are necessary for a managed or fiat system to yield a predictable value of money. That is to say, an operative commodity system works *independently* of governmental action or interference once the definitional step has been taken and the price of the commodity settled. The primary attractiveness of a commodity system lies in the limits that are imposed on the power of nation-states, on national governments, to debase the value of money by arbitrary increases in supply. The predictability that such a system guarantees is based on some assessment of the behavior of those large numbers who use and produce the monetary commodity. Such a system, if operating, insures against the unpredictability that stems from overt governmental interference, against the errors of explicit "management" attempts.

If a commodity standard is to operate within its own "principles," however, governments cannot intervene to offset the system's own internal discipline. Governments cannot allow inflation of the monetary base by encouraging the economizing of the monetary commodity, by legalizing and encouraging a fractional reserve base for note and deposit issue. Nor can central banks use the monetary commodity stocks in their possession to stabilize fluctuations in domestic price levels. Such attempts at management subvert the principles upon which the commodity system stands or falls, and they act to forestall, delay, and possibly to prevent the predictability that the system can embody if left to its own inherent logic, however uncompromising.

A commodity standard, as an automatically-adjusting monetary constitution, is effectively subverted once government so much as begins to superimpose management efforts. In the years between World War I and 1933, the so-called "international gold standard," and its United States component, was not a commodity standard at all. Because persons could convert dollars into gold, they were led to believe that the value of money was protected against wildly erratic fluctuations by the automatic adjustments that a commodity system embodied in principle. Individuals did not realize that national governments, and especially their central banks, had long since moved to interfere with the adjustment mechanism. As we look now at the pre-1933 monetary world, the breakdown is easy to understand; it is more difficult to understand why the system worked so long as it did without collapse. It was internally contradictory, a "managed commodity" system, which included neither the predictability inherent in the automatic adjustments of a commodity standard independently of governmental interference nor the predictability that direct governmental responsibility for monetary value might have induced.

What about the legitimate expectations of those persons

who entered into monetary agreements under this system? What entitlement did the holder of a gold-clause bond properly claim? What was his basic property right? If we impose rational wisdom about monetary theory on such contractors, the limits of such claimed property rights become apparent. The person who entered into a gold-clause contract for deferred payment should have recognized that, because gold was the defined commodity base for money issue, and because the system was not really based on the automatic adjustment process, almost anything might happen, including collapse of the whole monetary edifice. Pervasive uncertainty should have entered rationally into all deferred payment contractual negotiations.[26] The fact that such uncertainty was not more widespread suggests the quasi-fraudulent nature of the monetary framework. From our vantage point in history, the era seems to have been characterized by monumental confusion.

In such a setting, the parties who sought to make deferred payment agreements independently of the inherent uncertainty in existing monetary arrangements should have selected an agreed-on nonmonetary commodity or commodity bundle which could have served as a standard or index for deferred payment.[27] Governmental attempts at abrogation of such contracts would have clearly violated basic constitutional principles, at least as we have interpreted them. And we cannot infer from anything in the historical record that the Supreme Court would have, then or later, explicitly upheld either abrogation or prohibition of purely private

26. This was recognized, to an extent, in the opinion of Chief Justice Hughes, for the Court majority, in one of the gold-clause cases, where he noted that such contracts "have a congenital infirmity." (*Norman v. Baltimore and Ohio Railroad Co.*, 294 U.S. 240 at 307–308).

27. For a discussion of legal aspects of such contracts, see John P. Dawson and Will Coultrap, "Contracting by Reference to Price Indices," *Michigan Law Review*, 33 (March 1935), 685–705.

contracts in a designated nonmonetary commodity employed as a standard of value.[28]

VII. Governmental Restrictions on Private Contract

We now digress somewhat from our main discussion to the more general question concerning the possible bases for governmental power to restrict private contracts among persons, either explicitly or by refusal to enforce. The traditional economic argument for such possible governmental restrictions on voluntary trades or exchanges lies in the presence, or presumed presence, of third-party effects, of "externalities." If the carrying out of an agreement between parties affects the interests of third parties adversely, there may be grounds for governmental restrictions on the initial agreement. An "economic" theory of state or governmental action may be derived from this externalities approach, a derivation that is potentially consistent with contractual origins of governmental powers.[29] Examples are omnipresent in the discussion of applied economic policy, and notably in modern treatments of governmental policies relating to

28. It might be argued that obligations containing multiple currency options exercisable at the discretion of the obligee were logically equivalent to those in a nonmonetary commodity or index and that, on grounds similar to those advanced here, such obligations should have been held valid, even in the face of the gold-clause abrogations that had been upheld by the Court. Nussbaum takes this position with respect to a New York decision that upheld the refusal of the obligor to meet an obligee's claim for payment in Dutch guilder, although this was a specifically designated option under the terms. Since guilder had not been devalued, the obligee was, of course, rational in trying to exercise this option as opposed to his dollar claim. See Arthur Nussbaum, "Multiple Currency and Index Clauses," *University of Pennsylvania Law Review*, 83 (March 1936), 569–599. The decision under discussion is *City Bank Farmers Trust Co. v. Bethlehem Steel Co.* (244 App. Div. 634, 280 N.Y. Supp. 494).

29. See William J. Baumol, *Welfare Economics and the Theory of the State*, 2d ed. (Cambridge: Harvard University Press, 1965).

environmental quality. An individual purchaser of a new automobile cannot reach a voluntary agreement with his dealer for the removal of the pollution-control devices because of the alleged effect on overall environmental quality, which affects all persons and not only those who may be parties to the contract in question.

We need not either elaborate the analysis or extend the examples. We need to relate this externality basis for possible governmental restriction on private action to the monetary events of the 1930s. It is not surprising that, in a fractional reserve note and deposit issue system, the action of any person in converting or attempting to convert low-powered money (commercial bank deposits) into high-powered money (currency and/or gold), exerts potential external damage on others who hold and use money and claims to money in the economy. By the nature of a fractional reserve system, all persons are linked closely in an interdependent relationship which is highly vulnerable to behavioral changes. There has always been, and remains today, a possible contractual basis for governmental limitations on the set of institutional developments (from private unconstrained contracts) that produced the fractional reserve system of commercial banking. This is not, however, the position where the government of the 1930s chose to intervene. *Given* the institutions of fractional reserve note and deposit issue, there is an "economic" argument for granting government powers to inhibit the exercise of conversion privileges by private parties.

It is essential to distinguish between the government's exercise of its legitimate constitutional power to restrict trades or contracts among private persons in the presence of significant external effects and the extension of governmental power into an area where freedom of trade and contract had been hitherto protected by law. In the latter case there arises the issue of just compensation for those whose expectations are upset, whose property values are reduced or destroyed. These issues arise even when the new intrusion of governmental restriction may be justified on plausible contractarian grounds.

The freedom of persons to convert deposits and currency into gold at a long-established parity, and to make contracts in gold, was of value. As we have noted, there may have been a good contractarian argument for the suspension of convertibility because of the externalities inherent in any fractional reserve system. But since the suspension does destroy value should persons have been compensated for the losses suffered? This question is more easily raised than answered, but the closing off of valued options that could not have been exercised by all potential participants would not have been considered legal grounds for the payment of compensation.

We do not, however, need to answer this question here. We must, instead, go beyond the suspension of free convertibility and examine the possible externalities presented by the continued private ownership and contracting in gold, the basic monetary commodity. How would an individual's decision to hoard gold affect adversely the position of others in society? How would voluntary exchanges in gold impose external diseconomies? We have suggested an answer in our earlier discussion. If the suspension of convertibility does not fully destroy the mythology of the commodity as money, of gold as the standard of long-term value, individuals might have rejected national fiat money and they might have searched out ways to base exchanges on gold, even in the absence of governmental definition. A dual or even multiple money circulation might well have developed in 1933, and with this the destruction of some of the "public good" benefits of a single and unified monetary structure.

We find that, perhaps not surprisingly, welfare economics leads us to roughly the same point that our more general discussion did. There seems to be no externalities argument that can go beyond suspension of convertibility and restriction on ownership and trade in gold. There is no justification in the theorems of modern welfare economics for the government's failure to pay holders of gold and claims to gold at the new, rather than at the old, parity. Since the government would have lost nothing in the transaction, there could have been no adverse distributional consequences in a direct sense.

A somewhat different argument must, however, be applied to the gold-clause contracts between *private* parties. After devaluation, should the gold clauses have been honored, or was there an externalities argument for abrogation? As we have noted, in the absence of the anticipated increase in incomes and prices, debtors would have been distributionally harmed by enforcement of these clauses. Here we need to return specifically to our contractual setting for the selection of constitutional rules. Suppose that persons are behind a genuine "veil of ignorance" and/or uncertainty concerning their own roles as debtors or creditors in future periods. Are there any conditions under which they would choose to grant governmental authority to abrogate contracts among private persons? It seems plausible to suggest that some such authority might be granted for interference during periods of unforseen emergencies when the relative positions of debtors and creditors are exogenously and unexpectedly shifted. Both potential debtors and potential creditors might have conceptually agreed on such a grant of extraordinary authority to government, an authority which the abrogation of gold-clause contracts between private parties represents.[30] In this way, we can locate a more plausible argument for

30. In an analogous discussion, Mann introduces the "general doctrine of frustration," and he cites Viscount Simon in a British judgment to the effect that the event of a " 'wholly abnormal rise or fall in prices, a sudden depreciation of currency' . . . does not in itself affect the bargain but the true construction of the contract may show that the parties never agreed to be bound in a *fundamentally different situation* and that, accordingly, 'the contract ceases to bind at that point.' " (Italics supplied.) Mann points out that there is no United

this particular step in the whole New Deal package of monetary changes than we can for the confiscation of values by the calling in of gold coin and bullion at the old parity, and by the failure of government to honor the gold clause in *its own* contracts.

VIII. The Record of Government Failure

We emphasize that the monetary history of the United States is one of failure rather than success, failure of the government to accomplish, even tolerably, the objectives that seem to us to have been implicit in the constitutional delegation of monetary authority. The Great Depression merely dramatized this failure; the failure did not emerge suddenly and full blown in 1929. By 1933, the whole monetary structure was in disarray; basic changes in the rules were desperately required, and were seen to be required by almost everyone. The New Deal actions did not destroy a previously existing monetary system in any meaningful sense. The system was already in shambles. This point must be understood before any assessment of New Deal policies.

Ideally, the reconstruction of a new set of monetary rules should have commenced with the establishment of consensual agreement on the relative entitlements possessed by individuals and groups in the then-existing "monetary anarchy." But there was no time for attaining consensus, and for the genuinely constitutional deliberations that might have followed. The New Deal changes were imposed with little or no attention either to the distribution of entitlements or to the possible long-run consequences. There was, of course, distributional motivation behind the whole dollar devaluation package of legislation; one of the expressed purposes was

---

States decision on this point. See F. A. Mann, *The Legal Aspect of Money*, 3d ed. (Oxford: Clarendon Press, 1971), p. 120.

surely that of reducing the burden of debt and this could only have been accomplished by potentially damaging creditors, at least in one opportunity-cost sense. But creditors themselves could scarcely have expected to secure gains from continued banking crises, from continued erosion of confidence in the whole monetary and economic system. Had the New Deal changes been effective in securing the purposes for which they were intended, almost all individuals and groups would have benefited relative to the absence of such changes. The tragedy of the 1930s lies not in the New Deal's destruction of a long-standing, revered, and tolerably efficient monetary order. Such monetary order simply did not exist. The tragedy of the 1930s lies in the New Deal's own failure to "seize the day" and to institute those reforms that might have fulfilled the meaningful constitutional objectives for monetary order.

The pre-1933 monetary structure involved internal contradiction between its putative commodity-convertibility base and its vulnerability-exposure to misguided attempts at management in furtherance of domestic short-term goals. The New Deal's only prospect for change was toward more effective management, toward the establishment of an independent fiduciary standard, divorced from gold and from fixed foreign exchange parities. Unfortunately, the steps taken in this direction were fumbling and unsure ones, and the return to gold in January 1934, even in the limited sense of international convertibility only, became the device that seemed to justify carrying forward most of the unnecessary restrictions on individual freedom that the gold-based system had seemed to justify in the emergency.

As gold was gradually demythologized, there was less and less excuse for the continuation of the restrictions on private ownership of gold and on private trade in gold. As the economy recovered from the Depression, and as prosperity returned with World War II, the threat that individuals would, if given the opportunity, establish gold as a second circulating medium became remote indeed. The argument for continuing the restrictions in the late 1940s, 1950s and 1960s shifted to a wholly different base. The restrictions came to be supported on the grounds that it was necessary to economize on the world's gold stocks in order to facilitate and encourage international trading adjustments. To allow private trading and ownership in gold would have increased the so-called crisis of international liquidity. Had the Supreme Court accepted a challenge to the constitutionality of the prohibition of a gold-clause bond or of gold ownership in, say, the early 1960s, the continued restriction could have been justified within the same structure of argument as that used in 1935. The government's authority to continue the prohibitions first imposed in the 1930s could have been held to be required to prevent interference with the power to regulate the value of money, this time the value of the dollar with respect to "foreign coin."[31]

Since August 1971, or, more emphatically since December 1971, this basis for a legal justification of the gold restrictions could not have been sustained. Once the dollar was fully

---

31. Indeed it was precisely this argument that the government took in an effort to justify the continuation of the gold ownership restrictions in 1962. The authority of the government in this respect was successfully challenged in a single case in the Southern District of California. *United States v. Briddle and Mitchell,* 212 F. Supp. 584 (S.D. Cal. 1962). This decision was not appealed by the government under stipulation by the parties pursuant to a Supreme Court Rule. However, the decision was effectively reversed in the same district by another federal judge in 1965. *Pike and Brouwer v. United States,* 340 F. 2d 487 (9th Cir. 1965). In both of these cases, however, the government's authority was challenged, not in relation to the connection between the restrictions on gold dealings and the monetary authority of government but, instead, on the more legalistic point regarding the continued presence of "national emergency." In the context of modern legal history, any such challenge seems basically frivolous, since it seems clear that the Supreme Court would always defer to the Congress and the Executive in the definition of "national emergency." For a full discussion of the legal history, with citations, see Holzer, *op. cit.*

divorced from gold, in international adjustments as well as domestic transactions, gold has become basically equivalent to any other commodity. Unless it could have been shown that gold continued to carry with it important aspects of the mythology that surrounded it as a monetary metal, and, because of this, would greatly interfere with the monetary structure if freely allowed to circulate among persons, there would have been no constitutional basis for continuation of the restrictions on private ownership of gold, on private dealings in gold, and on the insertion of gold clauses in contracts. In more specific terms, the Supreme Court could not, in 1974, have upheld the validity of the Joint Resolution of June 5, 1933, on the same grounds that were used in reaching its 1935 judgment. The Court, of course, might have simply deferred to a congressional judgment, delegated to the Executive, in the determination of just what actions fall within the broad constitutional grant of monetary authority.

In this case, or in the absence of an opportunity for a clear judicial ruling, restrictions on the ownership of, and dealings in gold and, presumably, on the insertion of gold clauses in private contractual agreements, depend on the action of the Executive and the Congress. Although the President was given discretionary authority to remove the restrictions on gold ownership and exchange by legislation enacted in 1972, the Nixon Administration did not exercise this authority. Congress took additional action toward mandatory removal of the restrictions in late summer 1974, with the effective date set for December 31, 1974. President Gerald R. Ford signed the bill on August 14, 1974, although press reports indicated continuing administration misgivings.[32]

32. As late as December 1974, there were strong expressions of opposition to the removal of restrictions on gold ownership. Arthur Burns, Chairman

Over the early 1970s, the growing prospect that the restrictions on gold ownership would be removed undoubtedly exerted an influence on the free market price of gold. This price was surely higher than it would have been had the removal of the ownership restrictions not been anticipated. Over the long run, and especially if domestic inflation continues at the rates of the early 1970s or even accelerates, there might possibly be some development of institutions that would, in effect, utilize gold as a money competitive with the dollar, even if there is no governmental valuation of gold at all. In our view, this seems to be a highly unlikely prospect,[33] and it is certainly not one that would warrant reimposition of restrictions on private ownership of gold comparable to those imposed between 1933 and 1974.

IX. Inflation, Indexation, and the Monetary Authority of Government

By necessity, we look retrospectively at the monetary changes implemented by the New Deal in the 1930s. Considered solely as a subject of economic and legal history, those years hold their own intrinsic fascination. But the issues raised in that sequence of events contains relevance for the 1970s and 1980s, and it is this which motivates our inquiry. Begrudgingly, we are increasingly being forced to acknowledge that ours is an inflationary era, and that continued and possibly accelerating decline in the value of the dollar is the most likely course of economic events. Once this recogni-

---

of the Board of the Federal Reserve System, specifically suggested a six-month delay in the effective date of the legislation.

33. Our colleague Gordon Tullock dissents somewhat from our prediction here, and he places a considerably higher value on the probability that such a usage of gold would emerge, especially over a long period. He does not, of course, use this as an argument in support of restrictions on private property and contract.

tion is made and accepted, we are led to search for ways and means of adjusting to, of living with, inflation. If inflation is characteristic of our time, we can at least try to alleviate the burdens that it generates and to reduce the dislocation that it causes. (We shall not discuss why this setting has evolved or why a governmental policy aimed at stopping the inflation is not likely to emerge from political process. These would require separate papers, or books, in themselves.)

If the value of the monetary unit is to fall through time, how can individuals who seek to hold claims for deferred payment (potential creditors) protect themselves? How can individuals who seek current funds (potential debtors) secure these funds in sufficient quantity? How can long-term wage and salary contracts be negotiated in monetary units, in dollars, when the real value of these units is predicted to fall through time? How can real tax rates be set by Congress when tax brackets are denominated in money units?

During 1973 and 1974, there was mounting discussion of indexation as a means of adjusting to the uncertainty in the rate of inflation. Contracts may be drawn in a nonmonetary standard of value, in terms of some index for prices, in order to insure some meaningful relationship between the real terms on which an initial bargain is struck and the real terms on which the deferred obligation is to be met. Institutionally more and more wage and salary contracts contain cost-of-living or index clauses. Social Security benefits, and maximum payroll tax bases, were indexed in 1973. Senator James L. Buckley, with wide bipartisan support, introduced legislation in early 1974 that would require indexing of progressive income tax brackets, a step already taken in Canada, Denmark, and Holland. In May 1974, Senator Mike Mansfield called for the indexing of all wages and salaries. In July 1974, corporations began issuing long-term securities with interest rates linked to the short-term, Treasury-bill rate.

In this inflationary setting, the constitutional monetary authority of government may once again be challenged. At first glance, an index clause in any contract is similar in many respects to a gold clause. Each of these may represent an attempt by participants in an exchange involving a deferred payment to define their respective contractual obligations independently of the nominal monetary unit, the dollar in the United States. Having upheld the constitutional authority of the Congress to abrogate all gold clauses in contracts, would a modern Supreme Court uphold an attempt by the Congress, or the Executive, to abrogate or to prohibit the making of index contracts? This could become a critical economic and legal issue, especially if the government chooses to oppose the widespread proliferation of voluntary indexing institutions, a course of action that seems quite probable.[34]

There is, of course, a major legal distinction between possible governmental action in prohibiting future contracts between private parties and the abrogation of existing contracts. Congress might well be confirmed in its authority to prohibit all index contracts in the future but held to be powerless to abrogate existing index contracts. However, if indexing becomes more and more prevalent, the interest of the federal government in preventing future contracts may merge with its interest in abrogating then-existing indexed agreements. For our purposes, we may discuss these two actions at the same time.

The government's authority, if claimed here, would have

34. Legislative and/or administrative attempts to restrict indexation seem much more likely to occur in the late 1970s and 1980s than do any attempts to reimpose the restrictions on gold dealings that were eliminated in 1974. Index clauses, rather than gold clauses, may well become the focus of legal attention.

to be based on the interference exerted by such indexed contracts on the constitutional power to regulate the value of money, fully analogous to the gold-clause issues. There are differences as well as similarities present between the two situations, however. The situation confronted in 1935 was, in many respects, quite different from that which might confront the Court in 1976 with respect to index clauses. Abrogation of the gold clauses was upheld, at least in part, because gold was the accepted monetary commodity, because the contracts embodying these clauses were monetary contracts, because the ownership and trading in gold had previously been prohibited under an acknowledged constitutional authority. In all these respects, index clauses would be quite different. Deferred claims would be computed in a standard of value that remains independent from the medium of exchange, and there would be no side trading in the "index" as such.[35] The Supreme Court, should it decide to do so, could locate sufficient dissimilarities to reach a judgment contrary to those reached in 1935.

Nonetheless, there are also grounds upon which a court could confirm a governmental authority to abrogate index contracts and to prohibit further indexing. Widespread resort to such contracts would reduce the ability of the government to issue its own debt without itself adding index clauses.

And the employment-generating effect of unanticipated inflation would be weakened significantly in an economy where indexing was widespread. The government's ability to secure real resources for itself by expanding the money supply might also be hampered. Some persons might call

35. Mann places considerable emphasis on a 1957 French decision in which a contract for repayment in wheat was specifically held not to interfere with the monetary authority, a decision which, apparently, reversed a series of earlier judgments. Mann interprets this as validating gold clauses in France in the absence of legislation to the contrary. See F. A. Mann, *The Legal Aspect of Money,* 3d ed. (Oxford: Clarendon Press, 1971), pp. 152–53.

these effects interference with the regulation of the value of money.

A court should not, however, blindly support governmental authority to abrogate index contracts and to prohibit the making of these by private parties under the guise of the constitutional power to regulate the value of money. It should go behind this delegation of power in the Constitution and examine the possible objectives or purposes of governmental regulation of money's value. As our earlier discussion has suggested, regulation toward insuring greater predictability seems to be a plausibly meaningful interpretation of genuine constitutional agreement. If viewed in this light, there should be no grounds for upholding a governmental authority to prohibit or to abrogate index clause contracts. Such contracts are designed to insure, to debtors and creditors alike, predictability in the real value of the claims that are exchanged, predictability that would allow ordinary economic intercourse to proceed with minimum uncertainties about fluctuations in monetary values.

A question may be raised at this point, however, concerning the gold clauses that were abrogated in the 1930s. When contracts embodying these clauses were negotiated, were not debtors and creditors also attempting to insure predictability in the real values of the claims that were exchanged? This question requires a complex answer. In the first place, there seems to have been less-than-universal consciousness of an indexation basis for the inclusion of the gold clauses. In many such contracts, these clauses seemed to have been the result of rhetorical flourish on the part of bond lawyers, rather than consciously chosen instruments to insure stability in real value.[36] Second, the gold clauses were not, and could

---

36. The French gold clauses were abrogated in the 1930s on the basis of the argument that parties to contracts containing such clauses were not really aware of their existence. See Arthur Nussbaum, "Multiple Currency and Index Clauses," *University of Pennsylvania Law Review*, 84 (March 1936), 576.

not be, treated as instruments that would effectively divorce the real value of claims in particular contracts from fluctuations in the value of money. This follows, of course, from the fact that gold was the monetary commodity.

In a world of widespread indexing, the medium of exchange would continue to be the national fiduciary currency. The standard of value would be the designated index, which might, of course, vary from one contract to another. Obligations would always be paid in dollars, not in some other medium. The number of dollars required to meet any specific claim would, of course, depend on the change in the index, which would, in turn, depend on the change in the value of dollars in terms of some set of prices. Predictability in real value would replace predictability in nominal monetary units. But surely this would represent a change in the effective monetary constitution upon which all participants might possibly agree, save for the self-serving politicians and bureaucrats who might seek to maximize their prerogatives in resorting to the governmental money-creation powers.

Finally, and most importantly, index clauses would work quite differently from the gold clauses, despite the apparent similarities. The existence of index clauses would protect the parties against all shifts in monetary values and not just those reflected in a shift of a single price. This point may be emphasized by supposing that, in 1929, debt contracts should have contained index clauses rather than gold clauses. There was, between 1929 and 1933, roughly a 40 per cent decline in wholesale prices, and let us assume that this was the index used in the debt contracts. In this instance, the number of dollars that debtors would have had to pay creditors on obligations maturing in 1933 would have *fallen* by 40 per cent. The index clauses, in themselves, would have served to accomplish much of the debtor relief sought by the set of New Deal monetary measures. In sharp contrast, before devaluation the presence of a gold-clause did absolutely nothing to relieve the increased real burden of debt obligations. And, after devaluation, enforcement of the gold clauses might have subjected debtors to a doubly increased burden, one from the decline in prices and incomes, the other from the increase in the money price of gold. There could be no way in which index clauses could have generated such a burden on debtors in deflation.

The deflation in prices and in incomes that had occurred during the Great Depression was recognized by the majority of the Supreme Court in 1935. It seems clear from the argument in the gold-clause cases that these clauses would not have been abrogated if creditors could have demonstrated that they had suffered losses in real value due to inflation. There is nothing in the judgments of the Court that suggests an open-ended authority on the part of government to inflate the economy and to prohibit contracts that represent individuals' attempts to protect their real values against such governmental behavior.

IX. Conclusions

It would be pleasing to be able to conclude this paper with a statement to the effect that concerns such as those expressed in the preceding section are misplaced, and that modern governmental decision-makers will insure that inflation will be brought under control and depression prevented. Even if this optimistic prognosis cannot be made, it would be satisfactory if we could predict that a modern President, and Congress, would not deliberately seek to prohibit indexation or the abrogation of existing indexed contracts, that, if such attempts were made, a modern Supreme Court would declare such action to be unconstitutional, that the removal of the restrictions on gold ownership signals a permanent change in governmental attitude. Unfortunately, sober assessment allows no such prediction.[37] On monetary matters, present and future politicians do not seem likely to be more "enlightened" than past ones, and, if anything, they may be less so. The danger is great that modern governments, armed with the additional knowledge about the instrumental uses of money-creation powers and less fearful of challenging traditional mythologies about metallic-based money and balanced government budgets, may seek to secure an ever-increasing share of the economy's resources through continued inflationary financing.[38] To the extent that political decision-makers recognize the threat that widespread indexation poses for these objectives, they might well try to prevent such contracts and to abrogate existing ones.[39]

The modern Supreme Court has assumed powers of ultimate legislative authority.[40] The Court could protect persons from the ravages of government-created inflation. But it could just as readily do the opposite. The existence

37. An informed source has indicated to the authors that an official administration policy against indexation was laid down by Secretary of the Treasury George Schultz in late 1973. Further evidence of the government's opposition to indexation was provided in July 1974 by the reaction to Citicorp's highly successful issue of notes with flexible interest rates tied to the Treasury bill rate, setting the pattern for similar issues by other corporations. Arthur Burns, Chairman of the Federal Reserve Board, stated that such a note issue was contrary to the "public interest," and Congressman Wright Patman opened hearings on legislation possibly aimed at regulating such note issue.

38. As Robert Triffin emphasizes, "Man cannot unlearn the knowledge that man-made and man-controlled money gives to him." The mythology of gold could not be restored, even if this were desired. See Robert Triffin, "The Case for the Demonetization of Gold," *Lloyds Bank Review* (January 1974), p. 3.

39. As Mann notes, it is precisely during periods when index clauses would provide protection that governments would be most likely to prohibit and/or abrogate them. Cf., F. A. Mann, *The Legal Aspect of Money*, 3d ed. (Oxford: Clarendon Press, 1971), pp. 160–61.

40. "Constitutional anarchy" seems to be the most appropriate description of the legal disorder in which we live. For a more general discussion, see James M. Buchanan, *The Limits of Liberty: Between Anarchy and Leviathan* (Chicago: University of Chicago Press, in Press).

of legal precedent one way or the other would probably matter very little in the balance. We have moved a long way since 1935, when judges seemed to take "the law" more seriously.[41]

41. For those who dispute the dramatic shift in the attitude of the courts, we suggest that they read, Macklin Fleming, *The Price of Perfect Justice* (New York: Basic Books, 1974).

Gold, Money and the Law:
Comments

MILTON FRIEDMAN

Paul Snowden Russell Distinguished Service
Professor of Economics,
The University of Chicago

This fascinating paper raises a wide range of issues. I shall
comment on three: (1) the authors' account of the Great
Depression, the role of gold therein, and the subsequent
governmental measures taken with respect to gold; (2) the
authors' contention that the prohibition of the private ownership of gold and trade in gold might be justified by the
possibility that gold would be established as a second circulating medium; (3) the relation of gold-clause legislation to
indexation.

1. *The Great Depression.* The authors conclude their Section II, which is labeled "Before the New Deal," with the
statement: "A world monetary system built in such a way
on fractional reserves is vulnerable to a rapid succession
of bank failures and currency devaluations in major countries.
A combination of such confidence-shattering events *makes
it impossible* to satisfy the world-wide public demand for
currency and gold, given a low level of confidence in banks

and national currencies, without a substantial contraction in income. The accompanying reduction in the price level drastically increases the real burden of deferred payments, to the financial ruin of debtors" (my italics).

These sentences are correct for a hypothetical situation in which there is indeed a "low level of confidence" not only in banks but also "national currencies." But this condition was satisfied neither in the U.S. nor elsewhere from 1930 to 1933. As a result, the words "makes it impossible" cannot be justified in terms either of the actual course of events or of theoretical considerations.

In their next section, on the "New Deal Monetary Action," the authors assert: "In 1933, continued adherence to the international gold standard, as it was then operative, might have prevented or substantially delayed the internal monetary expansion that was so urgently required." And later, "The suspension of convertibility was a reasonable step."

Here again, I believe both assertions are questionable, except in the sense that they were self-fulfilling prophecies.

To turn to the first of the three quotations, there was in the United States no demand for gold prior to early 1933. The demand in 1933 itself arose only as a result of rumors, which proved correct, that the President-elect, Franklin D. Roosevelt, was planning to accept the advice of one of his key advisers, George Warren, to raise the price of gold in order to raise the prices of agricultural products. Had Roosevelt had no such intention and been able to make it credible that he had none, there almost certainly would have been no internal demand for gold. My impression is that there was similarly no internal run on gold in Great Britain or France or any other major country. Gold entered the picture, as the authors indicate earlier, as the mechanism for determining exchange rates and hence as an element in the transmission of deflation.

In the second place, bank failures outside the U.S., or currency devaluations in major countries, did not trigger the collapse of the world monetary system. On the contrary, the bank failures abroad (the Credit Anstalt failure, for example) and the British devaluation were both triggered by U.S. monetary policy plus the Smoot-Hawley tariff, which combined to produce balance of payments surpluses in the United States and an inflow of gold, which forced other countries to deflate or devalue. Austria (and other countries) deflated. Britain devalued. The bank failures and devaluations outside the U.S., rather than producing a collapse of the world monetary system, were simply manifestations of the collapse. And nowhere outside the United States was there a "world-wide public demand for currency and gold." The authors have generalized invalidly from the U.S. experience.

Bank failures on a wide scale affecting the public and producing a "public demand for currency" occurred predominantly in the U.S. They did not occur at all in Canada, Britain, or France, for example. In the United States, there had been many bank failures during the 1920's—roughly an average of one a day—but these had caused no serious problems and had produced no "public demand for currency" because they were mostly of small rural banks. For more than a year after the stock-market crash in 1929, there was no loss of confidence in banks, no attempt to convert deposits into currency—indeed the deposit/currency ratio actually rose a trifle, reaching a peak in October 1930. But then in the fall of 1930, a series of bank failures occurred culminating in the failure on December 11, 1930, of the Bank of the United States—the Franklin National of its time in the sense that it was the largest bank failure that had ever occurred in the U.S. up to that time, but very different in the sense that it was basically a "good" bank that could

and should have been saved.[1] This failure marked a change
in the character of the Depression. It then did become true
that there was a "public demand for currency" and that
the failure to satisfy this demand produced a monetary
collapse and the sharp subsequent contraction in U.S. income—but this was not the sequence elsewhere.

In the third place, and most important of all, while there
was in the U.S. "a low level of confidence in banks," there
was not "a low level of confidence in . . . national currencies,"
at least not until 1933 when rumors spread that Roosevelt
was planning to devalue.

This distinction between the confidence in banks and in
national currencies is crucial to the words that I italicized
in the first quote. It was in no way impossible to overcome
the low level of confidence in banks without a substantial
contraction of income, so long as confidence was maintained
in the national currency, as it was. I believe that Anna Schwartz
and I demonstrated that proposition pretty conclusively in
our *Monetary History.* Indeed it took an almost unbelievable
degree of ineptness and mismanagement on the part of the
Federal Reserve System to produce the catastrophic decline
in the quantity of money and in income that occurred from
1930 to 1933. Nothing in that process had the inevitability
that the authors implicitly attribute to it.

This rather different interpretation of the events that
preceded the 1933 banking holiday and suspension of convertibility is highly relevant to judging the two additional
statements of the authors that I quoted above.

In the absence of a rumored or known decision by Roosevelt
to devalue, I believe the authors are wrong in implying that
"continued adherence to the international gold standard"

1. It was not saved largely because it was one of only two major banks
in New York that were primarily in Jewish hands, and J. P. Morgan was unwilling,
apparently on that ground, to participate in a plan devised by the New York
Federal Reserve Bank to save the Bank of the United States.

would "have prevented or delayed the internal monetary
expansion that was so urgently required." Here again, Mrs.
Schwartz and I have argued the opposite. I hasten to add
that I believe that departing from gold in 1933 or earlier
was desirable given that the gold standard rules had been
so seriously departed from—in the 1920's by not expanding
the money supply as much as the gold standard rules called
for and in 1929–33 by contracting the money supply when
the gold standard rules called for expansion.[2] My point is
only that departing from gold was not necessary. In 1933,
the ratio of the U.S. gold stock to the quantity of money—at
the pre-existing parity of $20.67—was, I suspect, higher than
at any time in U.S. history. Certainly it was higher than
in 1929. And the flood of gold that came to the U.S. at
the $35 price makes it clear that that price was far too high.

The authors argue that "suspension of convertibility was
a reasonable step" because "the first requirement of any
policy was . . . to stop the attempted shifts into high-powered
money and to restore confidence in the system." I believe
this is wrong. The "first requirement" was rather to provide
additional high-powered money, which could readily have
been done without ending convertibility of gold at $20.67
an ounce, let alone $35 an ounce, and which would have
kept shifts into high powered money from reducing the
total quantity of money, and hence also would have restored
confidence by ending bank failures. Certainly a shrewd
observer at the time, Jacob Viner, was of the opinion that
"inflation," by which he meant an increase in prices and
in the quantity of money, could be produced without revalu-
ing gold or ending domestic convertibility.[3]

Of course, given that Roosevelt did intend to devalue under

2. See my "Real and Pseudo Gold Standards," reprinted in my *Dollars and
Deficits* (Englewood Cliffs, N.J.: Prentice-Hall, Inc., 1968), pp. 247–265.
3. Jacob Viner, *Balanced Deflation, Inflation, or More Depression* (Minneapolis:
University of Minnesota Press, April 1933).

Warren's tutelage, once that was known or even widely rumored, the situation was very different. At that stage, the only way to retain convertibility would have been to set a new price on gold at once and then continue convertibility at that price—a course of action, incidentally, that Buchanan and Tideman regard as potentially more effective to achieve what they regard as the justifiable objectives of the government action than the course of action that was actually followed.[4] But Roosevelt was unwilling to do that for two reasons. First, he wanted to experiment with changes in the price of gold because he had no definite idea at the outset of whether he did want to set a new fixed price and, if so, at what level. Second, he was unwilling to let private holders of gold profit from the rise in the price. The authors correctly stress this point, which in my opinion was in fact the only reason why the government tried to expropriate private holdings of gold.[5] The attempt did not fully succeed. There was no way of avoiding expropriation for gold certificates or for gold held in banks, but there is strong evidence that the greater part of the gold coin held outside banks was never turned in at the $20.67 price.[6]

In light of this historical background, I find it hard to enter fully in the hypothetical discussion whether the authors' "contractarian framework" can justify devaluation, the expropriation of private gold holdings at an artificial price,

4. The authors' analysis of this contention on page 36 is unsatisfactory. They refer to "a second and more direct force that could have been at work to spark an inflation in domestic prices and incomes," namely the increase in money holdings of the persons or firms who profited from the hypothetical devaluation without expropriation. However, there is nothing special about gold in this respect. In all cases, the creation of additional high-powered money by purchasing assets substitutes money for assets. Why should it be different whether the asset replaced is gold or government bonds?

5. The authors suggest this conclusion on page 25 but then seem to back away from it somewhat on page 26.

6. *Monetary History*, p. 464.

and the prohibition of private trading in gold. A definite course of policy was open first to the Fed then to F.D.R., and was urged at the time by knowledgeable and influential participants in the political process, that would have enabled the "contractarian" results to have been achieved without devaluation, expropriation, or prohibition. It was only necessary to increase high-powered money in the form of Federal Reserve notes and deposits.

Do they or we really want to take the position that any measures that could be justified by a "contractarian framework" under a set of hypothetical conditions are therefore justified even if those conditions do not prevail? Or perhaps more to the point, if those conditions (in this case, a demand for gold) are more or less deliberately produced by actions which themselves violate the "contractarian framework"?

The authors never really justify expropriation at an artificial price, except by analogy to the abrogating of the gold clause. Personally, I find it difficult to accept the analogy. The two cases seem to me quite different, one how to interpret the meaning and enforceability of a contract between two persons; the other, whether government is justified in commandeering private property without compensation. I have never been able to find any justification for the latter and do not find one in this paper.

2. *Suggested Justification of Prohibition.* As a result, I shall pass by the question of expropriation, take it for granted that devaluation occurred, and consider the hypothetical justification the authors offer for prohibiting the private ownership and trade in gold. This issue can be separated from expropriation because, as the authors point out, the government could have paid $35 an ounce for the gold it required private persons to turn over.

Say the authors, "It might be plausibly argued that, so long as gold was allowed privately to circulate, individuals would not retain confidence in using currency or bank

deposits, that confidence in the system would be restored
only if the private circulation of gold, independently of the
convertibility privilege, could have been prevented" (p. 32).
Or again: "A dual or even multiple money circulation might
well have developed in 1933, and with this the destruction
of some of the 'public good' benefits of a single and unified
monetary structure" (p. 55).

I find this justification so far-fetched as to be utterly
untenable. On the factual level, I do believe there was not
the slightest chance that a dual money system would have
developed in 1933 in the absence of the prohibition of private
ownership of gold and trading in gold. That did not happen
in other countries that did not enact similar prohibitions.
There were no signs of it at the time. There was no reason
for it to occur at the time because there was not then in
the U.S. and had not been at any time during the Depression
any public distrust of the national currency. The authors
have again confused loss of confidence in banks with loss
of confidence in a national currency—when the fact is that
loss of confidence in banks *enhanced* confidence in the national
currency. People fled from deposits to currency.

But second, a dual currency would not destroy to any
important extent the "public good" aspect of a national
money, if such a "public good" aspect existed. The U.S.
had precisely such a dual currency from 1861 to 1879 when
greenbacks and gold circulated side by side at a free market
exchange rate in the form both of literal gold and literal
greenbacks and of bank deposits denominated in gold and
in greenbacks. No doubt there were some accounting costs,
but I know of no historian of the period who has attributed
any substantial cost to the simultaneous co-existence of the
two moneys.

Third, on a purely theoretical level, I can envision a gold
private money replacing a paper governmental money under
only one set of circumstances: if the governmental monetary
policy is exceedingly erratic or highly inflationary. But in

that case, the existence of an alternative money that can replace the government paper is itself a good thing—may I say a "public good" rather than a "public bad." Consider Germany after World War I. Very late in the hyperinflation, many people started making contracts in terms of the dollar rather than the mark. Did this not reduce rather than exacerbate the harm that the hyperinflation did?

I conclude that the authors have provided no satisfactory justification on "contractarian" or any other grounds for either expropriation or prohibition.

3. *The Relation of Gold Clause Legislation to Indexation.* On a purely economic level, a contract to settle a future obligation in terms of a sum of money to be determined by the price of gold at that time seems indistinguishable from a contract to settle such an obligation in terms of the price of wheat or pork-bellies or of a share of U.S. Steel or of a market basket of goods. Granted that the widespread use of one or another of these devices may somehow affect the demand for other goods and services, including the demand for cash balances, and may affect the consequences of such government policies as financing deficits by newly created high-powered money. Granted also that a court might not be willing to enforce a gold-clause contract because the parties to it did not have in contemplation a situation such as actually developed. Presumably that might also apply to a contract stated in terms of "corn," if the meaning of "corn" changed drastically over time. But I do not see that either of these qualifications alters the essential economic identity among the various contracts.

On a legal level, the situation is apparently quite different because of the historical link between "gold" and "money" and because Congress has exercised its constitutional right to "coin money, regulate the value thereof, and of foreign coin" by specifying a price of gold in terms of the dollar. One admirable treatment of this difference, which I commend to the authors and to the participants at this conference,

is by Rita Hauser in two articles that deal with French
experience.[7] The first article, written in 1958, concluded
that, based on French experience, the abrogation of gold-
clause contracts did not imply any similar possible legal flaw
in indexed contracts. The second, written six years later,
reports the dismay among legal scholars over legislation
enacted under General Charles De Gaulle which abrogated
most existing index clauses and prohibited future clauses.
The article goes on to outline the exceptions explicitly
provided, and the attempts to find loopholes.

I do not myself see much likelihood that—all legal obstacles
aside—private individuals will in the foreseeable future find
it desirable to state contracts for future payment in terms
of gold. Gold has largely lost its role as international money
and with it any predictability of price in terms of "strong"
currencies or of purchasing power that it might have had.
It has never had the stability or predictability in terms of
purchasing power that "gold bugs" attribute to it. During
the nineteenth and early twentieth centuries its purchasing
power in terms of goods varied over a range of 3 to 1.
In the future, it will vary even more widely. It has now
become primarily a speculative commodity, with the added
feature that there is an overhanging stock in governmental
hands equal to something like 30 to 40 years' production.

But if individuals should choose to use gold, I share the
authors' judgment that it is hard to see how the courts could
declare such contracts invalid on the grounds of any close
link between gold and money. That link has surely been
broken and the removal of the prohibition on private ownership on gold is a highly visible and dramatic sign to that
effect.

7. "The Use of Index Clauses in Private Loans," and "Index Clauses in
Private Loans since the Advent of the Fifth French Republic in 1958," *The
American Journal of Comparative Law* 7, 1958: 350–365, and *13*, 1964: 606–611.

This means that, whatever may have been true in the past, as of the present all "indexed" contracts would appear to have the same status. If the courts nonetheless upheld Congressional acts declaring some sub-class of such contracts unenforceable, they presumably would have to do so on some basis other than the constitutional authority of Congress to regulate the value of money—despite the authors' ingenious attempts to find some basis in these terms. The lawyers here are far better qualified than I to say what such grounds might be. The experience of France and of Finland, both of which have tried to outlaw indexed contracts at various times, suggests that the issue is by no means purely hypothetical.

As the authors point out, the U.S. government has itself introduced index clauses into a wide range of future contracts, including Social Security benefits, pay of postal employees, etc. Many private wage contracts contain escalator clauses and so do many long-term purchase contracts for materials or for the construction of buildings or the like. The one area in the U.S. in which indexation has made least progress is the financial. The first big breakthrough came with the introduction by Citicorp of its so-called "floating notes." These link future payments to the then prevailing short-term rate of interest, which tends to move with inflation. The opposition to this innovation which was expressed by the thrift institutions, Congressman Wright Patman, and the Federal Reserve Board suggests that there is something less than a clear road ahead for indexation in financial contracts.

That road would be much easier if the federal government were itself to issue an indexed security, as there is much support for its doing. Such action is called for, as I have repeatedly stressed in other writings, by considerations of equity and representative government. In the present context, it would also have the very considerable advantage of reducing the possibility of legal challenges to private indexation.

Gold, Money and the Law: Comments

HARRY G. JOHNSON

Charles F. Grey Distinguished Service
Professor of Economics
The University of Chicago

Professors Buchanan and Tideman have produced an admirable paper that sorts out very usefully both the economic issues involved in the abrogation of the gold clauses in private and public contracts in 1933, and some of the constitutional issues as seen, not by a constitutional lawyer, but by the political theorist. One of the major difficulties in trying to clarify the issues in a specific case that is rooted in a bygone historical era with a bygone intellectual climate, however, is to establish and maintain the most useful perspective for learning the lessons of hindsight; and in this respect I find myself somewhat uneasy over three aspects of Buchanan's and Tideman's general approach, which I would have treated differently. In addition, I find it intellectually difficult to digest the idea that the cases involving gold-contract abrogations are relevant—or, at least, relevant without much more analysis in depth—to the question of indexation as a means for individual citizens to escape some of the major uncertainties of the current inflation.

Buchanan and Tideman remind us at several points that the constitutional theory applied by the Supreme Court has changed radically since 1933, in the direction of accepting any administration's own view of the powers it considers necessary to exercise its function of government. My reservations about their paper center on the fact that economic theory also evolves, sometimes extremely rapidly, and that its evolution is also shaped by institutional change in the broad sense.

In the first place, Buchanan and Tideman, in their excursion into the economic theory of constitution-making as it applies specifically to money, are at pains to argue that the constitutional power to regulate the value of money should be read as modified by explicit reference to the objective "so as to ensure predictability in this value." I cannot accept this as other than an anachronism. Historically, the colonial states had great difficulties with the provision of circulating medium, difficulties that continued and produced the National Banking Act, so that "regulation" could quite justifiably be deemed a clear end in itself.

More important, under the circumstances of the writing of the U.S. Constitution, and the prevailing state of belief about gold and silver, the value of a national money *was* its value in terms of precious metal. It was only towards the end of the nineteenth century, as a result of the "bimetallic" controversy about gold versus silver, that leading economists such as Alfred Marshall and Irving Fisher began explicitly to develop (1) the idea of purchasing power over goods in general as distinct from purchasing power over the precious metal, and (2) the idea of stability in this generalized purchasing power as the essential desirable characteristic of money and the objective of both the definition of money and of monetary management. The popular confusion between stability in terms of an internationally defined and accepted form of money, and stability in terms of purchasing power, it should be noted, remains with us today—both in the popular belief in gold as a hedge against inflation and in the efforts of the international monetary experts and officials to devise a reformed international monetary system based on fixed rates of exchange between national moneys.

Second, the use of a commodity, with certain well-explored physical characteristics, as money, represents a historical stage in the progression towards a pure credit money based on its acceptability as money and guaranteed in its acceptability by intelligent management. It is a half-way stage, in which the act of trust necessary for the organization of economic activity be specialization and generalized exchange (namely the willingness to break exchange down into: the exchange of something one values for something one does not want oneself and the subsequent exchange of the unwanted object for something wanted), is facilitated by the medium of exchange having some intrinsic value. This value may appear either to oneself or to the commonality of other people, so long as it limits the loss that may be incurred by accepting the medium in exchange for immediately useful objects. With the growth of general and unquestioning acceptance of the medium of exchange, that medium can be divorced from having a substance of intrinsic value.

The important point here is that the locking up of resources in giving intrinsic value to the medium of exchange is wasteful—a barren use of resources, which will be dispensed with as soon as and to the extent that general beliefs about what constitutes an acceptable exchange medium permit. The market will be seeking always to substitute virtually costless paper money and credit for commodity money with intrinsic value. The first of the main obstacles to this process, apart from the limitations of human ingenuity, is the value,

to the producers of monetary substances, of the public subsidy
inherent in tying together (1) acceptability and (2) monetary
substance as joint products in the provision of circulating
medium. On the other hand is the obstacle of public suspicion
of the process of creating something profitable (money) out
of nothing (paper and book entries) by the magical gesture
of passing private credit-worthiness over it. This is especially
so since that gesture sometimes gives rise to forms of
regulation that typically tied loans at favorable rates to
government and to certain classes of favored borrowers to
the acceptability-as-money feature.

The inherent wastefulness of commodity money and the
automatic pressures to find substitutes for it make me
doubtful of Buchanan's and Tideman's device of postulating
a pure commodity money system as a standard with which
to compare the monetary system as it existed in the pre-1933
period and as a basis for attributing inconsistency and
spuriousness to that system. One should not use as a standard
for comparison an alternative that is internally inconsistent
and contains the seeds of its own destruction.

In fact, it was experience with the self-replacing characteristics of the commodity money system that led to the
recognition of the need for control of the monetary system,
and with it, in a broader sense, of the need that "the regulation
of the value of money" had to be supplemented by an explicit
agreement on the objective of maintaining stability (less
restrictively, "predictability") is some sense. As the gold
standard development through the nineteenth and early
twentieth centuries, the main problem appeared to be "the
instability of credit" and the need for explicit acceptance
of a "lender-of-last-resort" responsibility on the part of the
central bank. In England, where the central bank was a
purely private institution, it took a series of crises to drive
home the need for responsibility of the central bank to the
monetary system as a whole. In the United States, the central

bank was established explicitly as a reserve bank for the private banks, in effect a lender of last resort, charged with the responsibility for maintaining "an elastic currency." However, this function was, until the mid-1930s, obnubilated by concern with the regulation of banking practices and the establishment of principles of "sound banking," including the usual vindictiveness towards unsuccessful innovators in the use of credit. In addition to this function, the Federal Reserve was charged with the general—but ill-specified and discretionary—responsibility for maintaining stability in the value of money.

This brings me to my third point of difference. Buchanan and Tideman seem to me to distract attention from the central broad constitutional issues by starting their story with the state of national disaster prevailing in 1933. The story, in my view, should instead concentrate on the failure of discretionary monetary management—both national and international—to guarantee stability of the value of money from 1929 on. Both nationally and internationally—or so I would maintain—what happened was that the initial steps in monetary management that were appropriate to counteract "the instability of credit" and that were fully within the experience and practices of the discretionary monetary managers and could have been effective, either were not taken or were taken with insufficient speed and determination. This resulted from a lack of leadership and determination, confusion on aims, and distraction from the central issue by side-issues. Once that failure had occurred—again, both nationally and internationally—institutional and legal restrictions on action, that in fair-weather times would have been manageable, became insuperable barriers to action. (An example is the difficulty caused by the Federal Reserve's rule that allowed only commercial paper to be used as a credit reserve backing for bank note issues in the face of a drying up of the supply of such paper.) Then the required action itself became something far outside the range of measures that could be contemplated by responsible and respectable orthodox financial opinion.

Viewed in this light, the policy actions of 1933 are merely incidents in this country's test case of discretionary versus automatic monetary management. The question of rules versus discretion, in its turn, is *the* central issue of democratic constitution-making for a liberal society, and it is an issue that divides contemporary theorists of the free society. The Chicago tradition has always in modern times emphasized rules. However, the nature of the rule has changed over time. The Henry Simons tradition emphasized, in line with older theorizing about the Federal Reserve, the imposition of a rule—price stability—regarding the objective to be sought by use of the discretionary powers of the monetary authority. On the other hand, the more recent Friedman tradition emphasizes the imposition of a rule—a given rate of monetary growth—replacing the discretionary use of the powers themselves.

The difference itself reflects in large part recognition of the fact that the authorities can always pay lip-service to the objective while taking policy actions directly contrary to its implementation. They can do this so long as they are entrusted with discretion in the choice of means for implementing any agreed end. The end, as Friedman's position illustrates, gives rise then to an insistence that to be effective a rule must govern what the authorities actually do and not merely their statement of goals. Part of the case for a monetary growth rule, a part whose importance is frequently overlooked, is that while the rule may not work ideally, its presence would provide an important benchmark for private calculation and correction, whereas discretion necessarily introduces an extra element of uncertainty and of speculation on the monetary authorities' behavior.

In contrast to the Chicago tradition, the European tradition, as reflected most notably in F. A. Hayek's *The Constitution of Liberty*, trusts the discretionary use of authority in the monetary sphere alone. This is for the well-known reason that money is the only economic good whose production cannot be entrusted to the market, not merely for the Buchanan-Tideman reason of externalities arising from the "public good" nature of the use of money associated with the requirement of acceptability, but also because money costs no real resources to produce, yet its usefulness and efficiency derive from the fact that it has a value in terms of goods and services in general.

The issue of rules versus discretion in monetary management is central, but it seems to me impossible to resolve, at least on the basis of current social theory. On the one hand, it seems inconceivable that any constitution-making convention, let alone any actual political state, would irrevocably forego a power—that of monetary management—the use of which might serve the general social good under certain conceivable circumstances. On the other hand, it seems equally impossible to devise a constitution under which those responsible for monetary management can be guaranteed to possess the requisite understanding of the social good, and of the monetary policy actions and techniques required to serve it, *together* with the independence of political action necessary to implement their responsibilities.

To return from this partial digression to the gold-clause issue, the economic issues seem, with the help of Buchanan and Tideman, fairly clear. The monetary authority allowed the economic collapse, and with it an unforeseeable transfer from debtors to creditors through a fall in prices. It then revalued gold, commandeered private holdings of gold to prevent private profits from its revaluation action, and abrogated the gold clauses in private contracts. The third action is to be distinguished from the second, because the second simply transferred unexpected profits from private gold holders to government—in the process indisputably counteracting, at least partially, the ostensibly reflationary intent of the revaluation—while the third eliminated the creation of unexpected private profits for some at the expense of creating unexpected private losses for others.

It is important to note that the distributional implications of the gold-clause abrogations vary according to the circumstances postulated. The post-1929 price deflation already involved unexpected gains to lenders on securities denominated in dollar terms, which would have been greatly increased by enforcement of the gold clauses. If prices had risen back to their earlier levels only, there would have been no redistribution to start with, but the enforcement of the gold clauses would have added unexpected gains and losses due to the rise in the dollar price of gold; and if prices had risen above their pre-Depression level in proportion to the increase in the price of gold, enforcement of the gold clauses would have meant no real redistribution among contractors protected by gold clauses, but redistribution from creditors to debtors on contracts not protected by gold clauses.

Under the circumstances, the abrogations dealt out very rough justice in an overall context undoubtedly favoring the creditors (apart from the non-negligible problem of bankruptcy and default). Note that in the case, characteristic of theorizing then and to a much greater extent now, of a small country in a large fully-employed world, the equivalent of a gold clause (e.g., denomination of a loan in a major foreign currency) would be a protection for creditors against arbitrarily-imposed losses resulting from devaluation of the currency of the country in which the loan was contracted.

How far does the case of the gold-clause abrogations serve to tell us something about indexation as a safeguard against inflation? That is really a problem for constitutional lawyers, not for an economist. However, it seems to me that the key to the matter, in common-sense terms, lies in the hazy distinctions between what can and what cannot be expected by a reasonable man, together with the presumed desirability of allowing the terms of private contracts to provide mutually beneficial protection against foreseeable and avoidable risks. From that point of view, indexation involves contracting on terms that protect a lender or seller against a general rise in prices that, because it has a high probability of benefiting the buyer or borrower in terms of his ability to pay higher money amounts of cost, imposes little if any additional real cost on him. In other words, indexation reduces the risk of real redistributions, by comparison with the process of contracting in money while attempting to foresee and guard against changes in the real value of money.

By contrast, a gold clause involves a strong element of uncertainty with regard to the purchasing power of gold over goods in general, the choice between gold and domestic money in a fixed exchange rate system favoring gold only to the extent that national monetary policy is more inflationary than that of the generality of countries (and hence devaluation-prone); or conversely, in a deflationary world, less deflationary than the generality of countries. In a world in which gold has no role as the monetary base of the international monetary system, however, and gold is simply another commodity, there is for the monetary authority nothing to be gained by prohibiting contracting in gold as contrasted with any other commodity—but also no significant private gain to be had from this freedom, apart from the possibly significant prospect of a return to gold at a higher monetary price.

Indexation does, however, raise a different question, noted by Buchanan and Tideman. Indexation, as a means of preventing real transfers of wealth between private citizens because of inflation, would, if applied to the operation of government itself (public borrowing, tax bracket definitions, pensions and Social Security, and so on), significantly reduce the scope for the use of inflation both as a covert means of increasing taxation without legislation, and as an excuse for intervening in market price-determination to achieve popularly-desired redistributions of income and wealth. Hence government has a vested interest in keeping the public ignorant about both inflation and the possibilities of automatic protection against it by indexation.

But to view government in this way is to get well outside the framework of the concept of a constitution—in which the purpose of government is to serve the public good—unless the subject is brought back within the fold by including as a constitutional goal the prevention of tyranny of the majority over the majority (or of various vocal minorities over the silent, uncomprehending majority).

In any case, it would seem unlikely that government aversion to indexation would be extended from the realm of government contracting with private citizens to a restriction on the freedom of private citizens to contract with one another on index terms. This seems especially so, since it would be patently inconsistent for government to try to combine its standard denial of responsibility for inflation (on the grounds that inflation is a natural disaster or one created by private greed) with a denial of the citizen's right to protect himself against losses for which the government takes no responsibility and which it proclaims itself powerless to prevent.

Gold, Money and the Law: Comments

RALPH K. WINTER, JR.

Professor of Law,
Yale University

I.

The threshold question for lawyers would seem to be whether the *Gold-Clause Cases*[1] were correctly decided, principally in the context of their abrogation in contracts between private individuals but also in obligations of the federal government itself.

Even a Supreme Court strongly bent on preserving liberty of contract, it seems to me, would have difficulty deciding the strictly private contract cases otherwise. Article I, Section 8 of the Constitution directs that "The Congress shall have Power . . . To coin Money, regulate the Value thereof, and of foreign Coin. . . ." While that does not in exact language accord a power to interfere with private contractual obligations, it also specifies no limitations. It thus seems the kind

---

1. *Norman v. Baltimore & Ohio R.R. Co.*, 294 U.S. 240 (1935); *Perry v. United States*, 294 U.S. 330 (1935).

of provision that the Supreme Court has frequently interpreted expansively so as to permit broad Congressional
discretion in the particular regulatory area. At the least,
for the reasons given by Buchanan and Tideman,[2] the
provision would seem to authorize the calling in of all gold
at the pre-devaluation price and thus stripping it of its
monetary functions. This interpretation is reinforced by the
apparently universal practice in other countries of entrusting
the central government with comprehensive monetary
powers, for a court faced with both a broad grant of power
in the domestic constitution and extensive foreign custom
as well would be understandably reluctant to attempt to
fashion restraints on its authority alone.

Existing precedent in the Supreme Court at that time,
moreover, revealed a steady expansion of the federal monetary power both as against the states and private parties.[3]
The *Legal Tender Cases*[4] established a power in Congress
to issue currency that creditors were obliged to accept in
satisfaction of debts which, when incurred, the parties had
reasonably anticipated would be paid in coin. These precedents, it would seem, were ample authority to support the
calling in of gold at pre-devaluation prices and a prohibition
on its domestic monetary use, again in light of the reasons
given by Buchanan and Tideman.[5]

This is a critical—in my view, dispositive—factor in
upholding the abrogation of gold clauses in private contracts,
even giving full weight to freedom of contract. Once gold
was stripped of its monetary functions, gold clauses ceased
to be a method of permitting private parties to adjust
underlying monetary obligations in light of subsequent
changes in the value of money.

2. Pp. 30–35.
3. See G. T. Dunne, MONETARY DECISIONS OF THE SUPREME COURT 1–83 (1959).
4. E.g., *Knox v. Lee*, 79 U.S. 457 (1871).
5. Pp. 31–32.

To be sure, if gold clauses were truly analogous to what we call indexing, considerations of freedom of contract would be at their strongest. Enhancing the ability of private parties to predict the future value of present contracts increases the incentive to engage in productive economic activity, and freedom of contract (assuming that to be a constitutionally validated policy) ought to prevail.

To the extent, however, that gold clauses were independent of the underlying contract, they seem more a wager on what action government will take in setting a price for, or prohibiting trade in, a particular commodity than an auxiliary and facilitating provision to a commercial contract. Considerations of freedom of contract in that case would lose much of their force. Bets on government action that are independent of any underlying contract may not encourage productive economic activity and may even create undesirable incentives to influence governmental action improperly. The virginity of freedom of contract may well thus be lost when the price of the particular commodity or the decision whether it can be traded—be it gold or widgets—is made by the government.

The Buchanan and Tideman analysis strongly supports the view that in 1934 the case for viewing gold clauses in the latter sense was powerful. Under the conditions existing then, creditors had not lost anything as a result of changes in the purchasing power of currency, the fear of which was the *raison d'etre* of gold clauses. It was a deflationary period and repayment of debts at face value in currency would thus not effectuate private contractual intent but would merely create windfall gains by government whim.

Freedom of contract simply does not call for such a result. What it calls for is the enforcement of the intentions of the parties and, while the enforcement of contractual language is generally the most suitable means to accomplish that, it is still only a means, not the ultimate end. Gold clauses were intended to protect creditors against changes in the value of money, not to create such windfalls. Enforce-

ment in 1934 would thus likely do more violence to genuine
notions of freedom of contract than abrogation, assuming
again the validity of government's prior actions toward gold.
Indeed, there were at the time precedents in which the Court
had indicated that a change in the value of money might
well affect the operative effect of other provisions in a
contract.[6]

One might plausibly argue, moreover, that the windfall
losses would result in a significant number of bankruptcies
and the very substantial transaction costs bankruptcy procedures impose. Government might, under such circumstances,
be thought to have the power to avoid those costs by
eliminating the windfall gains and losses.

As for the gold-clause obligation in government contracts,
Buchanan and Tideman argue persuasively that it presents
a different case. There the obligation is in no sense independent of the underlying contract because one of the contracting
parties set the commodity price and prohibited further
trading. Still, these cases are less interesting because of
government's unquestioned power to abrogate its contracts
through the outrageous yet simple device of not creating
a forum in which suits can be brought. Ultimately, Congress
did just that in these cases.

II.

Given that a plausible argument can be made in favor of
the *Gold-Clause Cases*, the next issue is the extent to which

6. *Willard v. Tayloe,* 75 U.S. 557 (1869) (denying the equitable remedy
of specific performance of an option to buy where the contract had been
executed at a time when gold and silver coin were the only lawful money
and the tender was made in legal tender greenbacks); see also *Effinger v.
Kenney,* 115 U.S. 566 (1885), (holding that a contract payble in Confederate
notes might be enforced after the war, recovery to be in the Union dollar
value of the notes at the time of execution).

See generally, Dawson and Cooper, *The Effect of Inflation on Private Contracts:
United States, 1861–1879,* 33 Mich. L. Rev. 852 (1935).

government may validly prevent private parties from employing contractual devices (among themselves or with the government) that will vary the obligations they create according to changes in the value of money. (For convenience's sake, I will hereafter refer to such devices as index clauses.)

Existing precedents are, to me at least, rather clear. The Supreme Court presently seems set on a course of decisions that, while as yet not explicitly rationalized, seem to distinguish between the power of government to regulate "individual" rights and its power where "economic" rights are concerned.[7] Under these decisions, for example, government is not permitted to regulate access to abortions because that regulation is said to intrude—curiously enough—on the right of privacy, and similar claims are being made as to all kinds of private consensual conduct.

On the other hand, the Court has made it clear that government has enormous discretion to regulate "economic" rights.[8] Under present precedents a serious challenge simply cannot be made to government's setting the price of any product or service, redistributing income, inflating or deflating the economy, nationalizing any industry, or, *a fortiori*, prohibiting index clauses. In each of these cases the Court's position is likely to be that the wisdom of the action is none of its business so long as any shred of "economic" purpose can be shown. Messrs. Buchanan and Tideman suggest, for example, that if private parties are permitted to index, the government's ability to fund its debt without itself agreeing to indexing would be lessened.[9] As a matter of prediction, I have little doubt that such a purpose would more than suffice to persuade the Court to uphold such indexing legislation

7. E.g., *Roe v. Wade*, 410 U.S. 113 (1973). See Emerson, *Nine Justices in Search of a Doctrine*, 64 Mich. L. Rev. 219 (1965).

8. E.g., *Day-Brite Lightening, Inc. v. Missouri*, 342 U.S. 421 (1952).

9. P. 64.

To predict how the Supreme Court will rule is not to justify the ruling. The distinction between "individual" rights and "economic" rights seems to me to be wholly artificial, except as a part of a personal value system. One may personally prize sexual gratification over economic gratification and thus regard the former as more essential to the exercise of individual rights than the latter. That is a matter of personal preference, however, and not of constitutional dimensions.

In truth, the distinction breaks down almost immediately as a constitutional rule. Those who argue that the abortion decisions can be justified as a facet of the right to do as one pleases with one's body are hard pressed to justify compulsory seatbelts as "economic" regulation. And those who would rubber stamp "economic" regulation have lately been nervously arguing that welfare regulations fall outside that category. One recent Supreme Court opinion, for example, tells us that "regulation of business or industry" differs from the "administration of public welfare assistance,"[10] while commentators have argued that state "economic regulation" is to be distinguished from state-imposed "distinctions that run against the poor."[11]

Not only does the distinction break down in both directions but it also fails to recognize that "economic" regulation is a potent device to infringe other rights, e.g., regulating the price of abortions.

To reject as artificial any constitutional distinction between "individual" and "economic" rights does not aid us in locating the parameters of Congressional authority to regulate index clauses, however. It does not settle the question of whether the interventionist judicial approach to abortion should be employed in measuring the constitutionality of restrictions

10. *Danbridge v. Williams*, 397 U.S. 471, at 485 (1970).
11. Karst and Horowitz, *Reitman v. Mulkey: A Telophase of Substantive Equal Protection* 1967 Supreme Court Review 39.

on "economic" activities or whether the courts ought to retreat in the area of "individual" rights to the "hands off" approach presently used in "economic" cases.

Buchanan and Tideman seem to me to be ambiguous on this issue largely because their idealized "choice settings" for determining the scope of constitutional restraints seem to be used to test the correctness of what the Supreme Court has done and or ought to do in the way of interpreting the Constitution, as well as to determine what ought to be in a hypothetical constitution. There is, of course, an enormous and obvious difference, but it is a major issue emerging in the legal literature. Although the problem of determining the legitimate sources of constitutional law has always been with us, John Rawl's book, *A Theory of Justice,* has led to a number of articles in legal journals which assume that the present Court should, if a "theory of justice" is established to its satisfaction, employ it in actual constitutional adjudication.

This assumption would, if widely accepted, represent a major institutional change in our system of government. For, while American history is marked by periods in which the Supreme Court appears to have undertaken to write the personal values of a majority of justices into the law, the legitimacy of their doing so has not, to my knowledge, ever gained prevailing acceptance. However constraining or difficult it may seem, I take the old-fashioned position that the Court is empowered to look only to theories of government and normative choices rooted in the Constitution itself and not to adopt on its own authority a new "theory of justice." This is largely a question of the extent to which notions of decentralized power call for limits on the Court's authority as well as limits on the powers of other branches.

There is no reason to believe that the Court is institutionally better qualified than the political branches to exercise unfettered discretion. The Court is not elected, and professional standards or methods do not justify a delegation of wholesale power to the Court to impose its view of what is the "good society" on the United States.

Much of the genius of the judicial process comes from the slow and deliberate elaboration of principles in a series of individual adjudications. Such a process insures that the full ramifications of a problem will be known and that decisions will be founded on adequate reflection and experience. If the judgments of the judicial branch are to be more enduring and entitled to a greater moral force than those of the political branches, piecemeal elaboration of general principles may be necessary. Creation of a good society, however, seems to me to call for a wholesale rather than a retail approach, one for which the judicial process is not well-suited.

The more complex the problem, moreover, the more likely are courts to resort to simple rules of thumb. One-man, one-vote, for example, is a rule stemming more from the inability of the judicial process to weight all the political imponderables involved than from its underlying merits as a principle of apportionment. Similarly, the "school equalization" cases,[12] if affirmed, might well have entailed an insistence on equality of per student dollar input as a means to remedy the nation's educational ills. All too often, in short, complex social problems leave the Court with a choice between "rubber stamping" the choices of other institutions or resorting to an inappropriate rule of thumb.

The history of the Court, moreover, suggests that it, like other branches of government, can become the captive of special interest groups. It is all very well to take potshots at a Court vainly attempting to distinguish between "individual" rights and "economic" rights. We should not, however, forget that a prior Court upheld the imposition of federal

12. See *San Antonio Ind. School District v. Rodriguez*, 411 U.S. 1 (1973).

regulations of a moral nature under the Commerce Clause while striking down economic regulation based on the same constitutional provision.[13]

The best way to limit the Court's authority is to create an ethos in the society and in the profession which insists that the norms the Court enforces and the authority it exercises be rooted in the Constitution. This is not a stump speech for "strict construction" or a demand for the elimination of judicial creativity. Judges must still draw inferences from "structure and relationship"[14] and elaborate and apply theories of government derived from the Constitution in concrete cases. They must still seek to accommodate and reconcile conflicting constitutional values and permit growth and change in the light of history and experience. But the structure and relationship are the structure and relationship established by the Constitution, and the norms to be elaborated must be rooted in that Constitution rather than in a particular judge's personal value structure.

For my own part, this means the Court was likely wrong in the abortion decision and right in recognizing broad congressional authority to limit economic freedom. Neither result truly conforms to my preferences when presented as legislative propositions. Still the Constitution is not a charter forbidding legislative mistake. Nor is it a charter authorizing the Supreme Court of the United States to fashion its own version of the good society.

III.

The final issue, then, is what the "good" constitution ought say about government power over money—or, put another way—to what extent should the Constitution be amended.

13. Compare *Champion v Ames*, 188 U.S. 321 (1903), with *United States v. E. C. Knight Co.*, 156 U.S. 1 (1895).
14. C. Black, STRUCTURE AND RELATIONSHIP IN CONSTITUTIONAL LAW (1969).

The basic function of a Constitution in a democratic society is to create the power to govern and to surround it with a structure that is consistent with the long-run survival of relative freedom. The structure should be constitutionally inhibited in two ways. First, procedural regularity should govern the workings of each branch or agency established. Second, power should be decentralized to a feasible degree through a system of checks and balances and/or a federal structure.

Limitations on the substantive powers of government—e.g., to subsidize an industry—should be imposed only in very compelling circumstances. We simply cannot comprehend either the policy perspectives, value preferences, or learning of future generations, and overeagerness in imposing "permanent" substantive policies is more likely to limit the endurance of the Constitution than to prolong the effective life of the policies themselves. To be sure, law can add to the moral as well as to the coercive force of a policy. In the long run, however, it cannot effectuate policies that have lost their persuasive hold on society, and it will only discredit itself in the attempt.

Restraints on substantive regulatory power can, however, be justified where certain conditions (herein greatly simplified) are met. First, the issue must be of fundamental dimensions, one, indeed, that goes to the foundations of the society itself. Second, there should be some demonstration that the political branches of government may be less neutral than the courts in enforcing the rights or obligations involved. This means either that the issue is somehow unsuitable to majority rule or that institutional pressures within the political branches may be thought to be ill-suited to carrying out the particular policy. Third, the issue should be of the kind that courts can successfully resolve by judicial decree.

Finally, viewed from some "choice setting," the issue should appear to be resolvable in terms of a principle of general acceptability, which principle might well be abandoned if subject to the will of transient political majorities in particular political contexts. Thus, both Rawls and Buchanan and Tideman pose "choice settings" that compel us to view ourselves as being as likely to be in a minority as in the majority.[15] This compels us to focus on the principle rather than on a particular political context in which we, being in power rather than out, might gain from its abandonment. Such an approach to constitution-making is essential because all would agree that transient political majorities do not always reflect the considered will of the society or maximize that majority's options. A majority may well prefer a general principle only when assured that it will be neutrally and generally applied over a long period, since in the absence of the assurance that it will be applied when it is out of power, it would be rational for the majority to press its temporary advantage.

Consider two examples, federalism and freedom of speech. Federalism is an issue of fundamental importance to those who value the decentralization of power. It is not so clear, however, that the political branches will not, in their interaction, reflect in significant ways such considerations. Both Representatives and Senators are elected within the state units, as are electors in the electoral college. This tends to increase the sensitivity of the Presidency and Congress to State claims. Beyond that, the very existence of states as going governmental entities itself poses a check upon the federal government. Political entities such as states develop an institutional life of their own—and institutional claims that they press energetically. To be sure, states' rights have been eroded, but those who work in law and economics are often so concerned with federal regulation that they tend to ignore the fact that the vast preponderance of law

15. P. 42.

governing our lives is state law while federal law tends to be limited and interstitial. Economic regulation, moreover, really is a national problem and those who resist the exercise of federal power in this area are usually also opposed to its exercise at the state level. A free market, after all, calls for freedom from all regulation, whether state or federal.

Beyond that, and more critical, there really are not rules suitable for judicial enforcement that distinguish state responsibilities from federal. What is involved are a series of presumptions rather than prohibitions, presumptions involving political judgments as to the seriousness of the issues involved, the validity of the proposed policies, the strength of public opinion, and the likely effect on state power.

Finally, I question whether a society, from whatever "choice setting," would attempt to map out in great detail substantive rules to govern our federalism. All of the difficulties of limited foresight inhere in such a task and I should think that one would opt instead for structural checks.

Contrast this with the problem of freedom of expression. It is, to be sure, an issue of significance, particularly where political speech is involved, for the existence of an electoral system has implications calling for the freedom to press one's political claims.

The lack of neutrality in other branches is also at its greatest where political speech is involved. Incumbents have every incentive to be decidedly partial in the regulation of political speech, and the need for a relatively more neutral institution is great. Nothing demonstrates this so much as the recent "campaign reform" legislation, which bears the stamp of incumbent self-interest in virtually every provision.

Judicial enforcement, moreover, seems appropriate for reasons other than relative neutrality. Interpretation of the First Amendment does, of course, call for the exercise of judgment and the weighing of conflicting considerations. Still, what is ultimately involved is nay-saying, a considerably easier task to carry out than the reordering of a legislative apportionment or of an educational system.

Finally, as Professor Friedman has noted, a society will, from a "choice setting" in which no one knows in advance whether he will be in power or out, be likely to opt to protect speech.[16]

In the case of money, Buchanan and Tideman seem to me correct in concluding that constitutional provisions regulating government's power over money can be justified. The power to regulate money is the power to wreck an economy and the society it supports. As Buchanan and Tideman demonstrate, moreover, the political branches of government may have institutional interests in abusing that power, the present example under discussion being the New Deal's profiting from the devaluation of 1934. Other examples are the deliberate use of inflation as a devious form of taxation benefiting the government. From a "choice setting" perspective, similarly, it would appear that at the very least, a society could agree on commanding predictability in a way that might permit individuals to avoid government's taking advantage of their fixed arrangements.

Whether further substantive limitations on the monetary power ought to be imposed or whether a particular method of assuring predictability ought to be adopted, however, runs into the problem of assigning tasks to courts that are judically "undoable." If there are relatively fixed and identifiable rules as to the regulation of money and if these satisfy the other criteria, then perhaps a comprehensive constitutional amendment would be desirable. That seems not to be the case, however. Similarly, predictability is not something that can simply be legislated, because, if for no other reason, no index is universally satisfactory.

The solution, I believe, is to permit contracting parties

16. M. Friedman, CAPITALISM AND FREEDOM 51–52 (1962).

to "legislate" the amount of predictability they prefer and to work out details such as the proper index themselves. We should, in short, give constitutional protection to index clauses between private contracting parties. Additionally, we might consider compelling the government to include index clauses in its contracts, both to decrease incentives to inflate and to reduce the likelihood that government will, because of difficulties in borrowing, seek to subvert or circumvent the protection given to private index clauses.

Gold, Money and the Law: Comments

GERALD T. DUNNE

Professor of Law,
St. Louis University

Since there is no prompting of pride and no subtle tilt toward reflex approval quite like that which besets a commentator who encounters a reference to his own published efforts in a work on which his own views are requested, an early assertion of disagreement might be in order both to clear the air and sharpen the focus of discussion.

For notwithstanding my admiration for the authors' erudite and arresting analysis of the law and money, I must note my dissent from the interrelated theses that "an accelerating decline in the value of the dollar is the most likely course of events" and that a "government policy aimed at stopping inflation is not likely to emerge from the political process."[1] On the contrary, I disagree and respectfully insist that the time is ripe for a statutory resoltuion of our monetary crisis.

I must even confess a sense of shock to encounter the names of Buchanan and Tideman brigaded with the Keyne-

1. Buchanan and Tideman, pp. 61–62.

sian *epigoni* in an insistence that our present state of affairs readily admits neither of diagnosis nor therapy but instead must be borne for a while as a kind of inexorable *Zeitgeist*.[2] Yet as the great baron himself asserted in a prelude to his famous truism on death and the long run: "Economists set themselves too easy, too useless a task to tell us in the long run when the storm is long past, the ocean is flat again."[3]

Having duly disagreed, let me hasten to note some points of concurrence with the principal work—especially an agreement as to the profound revolution in constitutional attitudes that has taken place, the extraordinary mutation that has overtaken the judicial function. On the one hand, it guarantees virtually reflex approval of any legislative act, especially one of economic regulation, and on the other hand, it offers a new meta-legislation, itself the product of the new judicial avant-garde. The late Justice John Marshall Harlan summed this up as the twin doctrines (1) that no social ill which besets the nation cannot find its cure in some constitutional "principle" and (2) that the courts should take the lead in promoting reform when the other branches of government fail to act.[4]

Thus, I would agree that the drift of attitudes is such since the *Gold-Clause Cases* that re-argument in 1974 before the Supreme Court would result in a 9–0, not 5–4, judgment of that tribunal. More than this, as I read the fall-out from the *American Writing Paper* case,[5] I must likewise conclude that a statute outlawing all index-linked agreements would be sustained out of hand, perhaps pursuant to two dicta of Justice Robert Jackson: his troubled statement that, under

2. See, e.g., Paul Samuelson, Worldwide Stagflation, "*Morgan Guaranty Survey*," June 1974.

3. John Maynard Keynes, *A Tract on Monetary Reform*, 1923, 167–168.

4. Justice John Marshall Harlan (dissenting) in *Reynolds v. Sims*, U.S. 533, 1964, 377.

5. *Holyoke Water Power Co. v. American Writing Paper*, U.S. 34, 1937, 300.

the new dispensation, the Justices "have no function but to stamp this Act "OK"; and his Godkin lecture warning that as to the ravages of inflation, "No protection can be expected . . . from the Judiciary. The people must guard against these dangers at the polls."[6]

Let me add that I am far from unsympathetic to indexation, but, as Professor Friedman himself notes, "it is far better to have no inflation and no escalation clauses."[7] In any event, the completely indexed society seems to me not unlike that mythical island of Anatole France where each half of the population lived very well by taking in the washing of the other half.

However, let me turn to explore the possibilities of the other half of the new judicial avant-guardism of "have theory—want case." On the basis of Justice Macklin Fleming's latest pronouncement,[8] I might venture to put forward (partly in whimsy, but only partly) the hypothesis of Marshall McLuhan and Alfred Korzybski, that money is really a language,[9] as a basis for concluding that our constitutional quest should be to seek the basic guardianship of money under the cover of the First Amendment rather than the Fifth.

We are at a critical intersection today between a recollection of the past and a concern for the future that will, I believe, permit the drafting of a new monetary constitution. It should

6. Note to Chief Justice Stone, May 25, 1942, in Alpheus T. Mason, *Harlan Fiske Stone,* 1968, 595; Robert H. Jackson, *The Supreme Court in the American System of Government,* 1955, p. 59.

7. Milton Friedman, "Using Escalators to Fight Inflation," *Fortune,* July, 1974, p. 94.

8. *People v. Murphy,* voiding the parole limitation provision of California's heroin constraint statute, *Federal Reporter,* October 14, 1974, 7; Justice Macklin Fleming's anti-activist polemic, *The Price of Perfect Justice,* is favorably noted in Buchanan and Tideman, note 41.

9. See Marshall McLuhan, *Understanding Media,* 1964, 131–144; Alfred Korzybski, *Science and Sanity* (3rd ed.), 1948, 76–77, 519, 549.

perhaps be called the Anti-Inflation Act of 1975 and enact
the *Monetary History of the United States* into law in much
the same manner as the Employment Act of 1946 codified
the *General Theory of Employment, Interest, and Money*. As you
know, the *History* is a formidably rigorous development of
three theses:

1. Changes in the money supply are closely associated with
   changes in economic activity, income, and prices.
2. The interrelation between monetary and economic
   change is highly stable.
3. Monetary changes are often of independent origin; they
   do not simply reflect changes in economic activity.[10]

A flood of scorn has poured out over these ideas all the
way from Kings College in the old Cambridge to Harvard
College in the new. And this cacophony of criticism—that
the monetarist thesis is superficial, implausible, simplistic,
and only a partial guide to the ambiguities it seeks to
explain[11]—could be said to reflect the thesis' strongest
political credentials. The thesis admirably fills Henry Simons'
norms for a monetary constitution, as set out in his classic
"Rules versus Authority in Monetary Policy:"[12] (1) non-
reliance on discretionary (dictatorial, arbitrary) action of an
independent monetary authority, (2) simplicity and defi-
niteness, and (3) clarity and rationality sufficient to regiment
strong public sentiment around a new 'religion of money'
as an insulation against further tinkering.[13]

All this is prelude to the central point, which is "to design
and establish, with the greatest intelligence, a monetary system

10. Milton Friedman and Anna Schwartz, *A Monetary History of the United States, 1867–1960*, 1963, 676.
11. Quoted in *Burroughs Clearing House*, May, 1974, p. 23.
12. Henry Simons, "Rules versus Authority in Monetary Policy," *Journal of Political Economy*, 44 (1936), 1, reprinted in *Reading in Monetary Theory*, 1951, p. 337. Citations herein are to the latter source.
13. *Ibid.*, pp. 340–341.

good enough so that, hereafter, we may hold to it irrational-ly—on faith—as a religion, if you please."[14] The paradox, however, is more apparent than real, for, as Korzybski says, "the reality behind money is doctrinal, 'mental,' and one of the most precious characteristics of the human race."[15]

Remarkably, in terms of the biography of an idea, we could well focus on Congress and begin our review by recalling the Spring of 1952 when the then Senator Paul Douglas enlivened a Congressional hearing and made wire service news by emptying a pitcher of water into a much smaller glass. His action was no accident. On the contrary, it was intended as a scientific demonstration, not in physics but in monetary economics, for the benefit of the incumbent Secretary of the Treasury, John W. Snyder:

> Mr. Secretary (said Douglas), if you force the Federal Reserve System to purchase additional large quantities of government bonds, thus expanding bank reserves, thus expanding credit, the task of trying to prevent prices from rising . . . will be just as futile as when I fill this glass of water and keep pouring it in, and try to mop the overflow with a pocket handkerchief.[16]

Douglas had undertaken the demonstration because of an apprehension that the Federal Reserve, at the Treasury's behest, might resume purchases of government bonds in amounts necessary to keep their market price even with their stated par value and their market yield level even with their coupon rate. Introduced as a matter of wartime finance and thereafter converted to an instrument of peacetime stabiliza-tion, the Federal Reserve's support policy held down interest rates not only on the largest government debt in history, but on all loans throughout the entire economy. It did so,

14. *Ibid.*, p. 350.

15. Korzybski, *op. cit.* note 9, pp. 76–77.

16. Hearings, "Monetary Policy and Management of the Public Debt," Subcommittee on General Credit Control and Debt Management, Joint Economic Committee, 82nd Congress, 2nd Session, March 30–31, 1952, pp. 22–23.

however, at the cost of a virtually open-ended money supply. Largely as a consequence of Senator Douglas' prodding over inflationary consequences, the policy had come to an end a little more than a year before the hearing in the famous Accord of March 1951, but its resumption was always possible, and the contingency of a return to the *status quo ante* had prompted the hydrostatic experiment.

Thereafter, the gyrations of Federal Reserve behavior left a mark on Congressional documents. Following the rapid expansion and non-growth of money in the crunch year of 1966, the Joint Economic Committee urged a policy of moderate and relatively steady increase in the money supply "generally within a range of 3–5 percent a year,"[17] with the minority members suggesting a range of 2–4 percent. In 1968 the Committee broadened the band of suggested growth rates to 2–6 percent.[18] The crunch of 1969 apparently aborted further general suggestion, but in the 1970 report, the minority members suggested a band of 2–3 percent with a "safety valve" lockup of 2–6 percent.[19]

Indeed notwithstanding the lapse of congressional suggestion and the complaint of Congressman Henry Reuss' Committee that the dialogue of the Federal Reserve and the Joint Economic Committee might well be "conducted in Urdu on one side and Swahili in the other",[20] the grammar and syntax of monetarism has been an especially receptive vehicle for congressional perception of monetary policy, as a comment by Senator William Proxmire has evidenced:

> What they (the Federal Reserve) are doing is contributing to the inflation. In six months, as I understand it, December through

17. Joint Economic Committee, *1967 Economic Report*, 1967, p. 14.
18. "Standards for Guiding Monetary Action," Hearings and Report, Joint Economic Committee, 1968, p. 17.
19. Joint Economic Committee, *1970 Economic Report*, 1971.
20. "Supplementary Views of Representative Reuss," *1968 Joint Economic Report*.

May, the money supply increased by more than 8 percent. What kind of restraint is that? Burns is like Nixon—he talks a good fight. But when you see what the bottom line is, what's happened to the money supply, or what's happened to spending—the [former] president talked about his restraint in fiscal policy and his prudent budget—you see the biggest increase in history in the money supply in a peacetime year, percentage-wise as well as dollar wise.

The Fed has been saying that this double-digit inflation puts our society in jeopardy . . . well, if that's the case, we're not doing anything about it except aggravating it. I understand it would be very difficult to have restrained the money supply to say, a 4% growth. But neither the Fed nor the rest of the economic managers in our country are doing one single thing about it except talking about it and letting the economy go the other way.[21]

Against this background, however, and in the post-Watergate mood, a Congress restive to regain its constitutional prerogatives might well wish to command where it has previously counseled. And the Congressional as well as the popular mood might be prodded by public examination of the Federal Reserve's statements that have been citing the very theoretical standards that the Federal Reserve has failed to honor in practice.

Interestingly, in its current effort to sweep all demand-depository institutions into *de facto* Federal Reserve membership, the Fed has pleaded that "the Federal Reserve's control over bank reserves (and the money supply) has been eroding."[22] And, ironically, the more recent proposals, squinting at socialized collection payments under Federal Reserve proprietorship,[23] have also been based on an appeal to enhanced manageability of the money supply.[24]

21. R. L. Miller (Ed.) "A Conversation with Senator Proxmire," *The Bankers Magazine,* 25, (Autumn, 1974).
22. "Uniform Reserve Requirements," *Federal Reserve Bulletin,* 1974, p. 168.
23. 38 Federal Register 32952, (1973).
24. See "Fed's Changes in Regulation J are Pointed to Electronic Transfers," *American Banker,* December 1, 1972.

And so I have drafted some proposed legislation, a copy of which is appended. Basically, this is a proposal both to rap the Federal Reserve's knuckles and to clip its wings. It is in two parts, the first a revision of the Employment Act to enumerate price stability as a formal objective of national economic policy. This is a purely symbolic gesture, but as Justice Holmes said, "We live by symbols."[25] (Almost thrity years ago Henry Simons also said that we need a declaration against further price increases.) The second part requires that, after recommendations of the President and the Joint Economic Committee are in hand, Congress set the annual rate of increase in the money supply (narrowly defined). This part would also revise the incredibly elastic provisions of the Federal Reserve Act—which presently tell the Federal Open Market Committee simply to go forth and do good—to mesh with the revised Employment Act.

It must be realized that any statutory solution does violence to one of the most revered symbols of our current monetary constitution, the independence of the Federal Reserve. And here let me dissociate my view from Professor Friedman's conclusion that, in view of the Fed's record of monetary contraction and expansion, the country would have been better off without it. On the contrary, I think it self-evident that the Federal Reserve is a valuable national institution worthy of public understanding and support, and it will not be improved by explicit politicization or the hectoring of the General Accounting Office.

But the facts of the matter are, of course, that the emperor has no clothes, nor the Federal Reserve any independence—at least for the past third of a century. Surely the evidence is clear that the replacement of the "Peg" of 1941—with its pledged support of government bonds, for the accord of 1951, pledging support of such bonds' markets—merely

25. Holmes, *Occasional Speeches*, 1962, p. 134.

exchanged King Log for King Stock, and the sinister upward trajectories of the price levels, the money supply, and the Fed's holdings of government securities express three intimately related consequences.

But if we can assume a statutory commitment to price stability plus a demise of the accord of 1951, would it not be possible to remit the Fed to true independence and discretion rather than merely to symbolic existence? The answer I fear must be in the negative, for as James Meigs has well said, there is no more reason why the United States monetary policy should be handled by an independent Fed than its military policy be handled by an independent Pentagon.[26] Beyond this, however, the Fed itself appears to have undergone a fundamental mutation and appears to be exhibiting bureaucratic symptoms to which it had long appeared immune.

First there is a kind of hubris, a disposition to put its hand to many things it cannot do—aid housing, influence interest rates, safeguard the balance of payments, etc., etc.—in preference to the one thing it can do: control the money supply.

Second is a symptomatology whereby the new Fed differs in character, outlook, and attitude from the old. The most conspicuous aspect of this is the way it has conformed to Parkinson's Law by expanding its staff as its service constituency shriveled.[27]

The Fed's response to the declining constituency is not without interest—compulsory membership disingenuously brought forward as uniform reserve requirements and justi-

26. Hearings, "The Credit Crunch and Reform of Financial Institutions," House Banking and Currency Committee, 93rd Cong., 1st Sess., Sept., 1973, p. 261; See also Sayers, *Modern Banking* (1964) 65–81.

27. Between 1969 and 1973, the Fed's primary service constituency-state member banks, fell from 1313 to 1092 while Reserve Bank employment expanded from 20,128 to 26,655.

fied on the basis of the seldom-used money-supply control
as an instrument of monetary policy.[28] Along much the same
line has been the use of the Bank Holding Company Act
as a wedge to System membership, notwithstanding earlier
assurances to the contrary and settled Congressional policy
that the statute would be optional in that regard.[29] Finally
there has been the institutional self-aggrandizement of bringing the electronic revolution to the national collection-
payment mechanism.[30]

So the hour seems ripe for the recasting of our monetary
constitution. Indeed, if I were forced to historical analogies,
I would say that in terms of the massiveness and seeming
insolubility of the problem, the present moment in our affairs
is not unlike that which faced Senator Douglas at the
demolition of the "Peg," or that which faced the draftsmen
of the Resumption Act of 1879 or of the Federal Reserve
Act of 1913. Senator John Sherman wrote a decade before
the enactment of the first statute: "I have my own theories
on the mode of redemption, but the process is a very hard
one and will endanger the popularity of any man or administration compelled to adopt it."[31]

In terms of timing tactics, Woodrow Wilson, on the eve
of the legislative struggle for the Federal Reserve Act, said
everything there was to say: "We shall never find the time
. . . because whenever such action is contemplated the same
obstructions will arise."[32]

Perhaps the best epigraph of all can be found in the
copperplate prose of Wilson's first inaugural:

28. See note 23 *infra.*
29. See Jerome Shay, "The Emerging Federal Reserve Primacy in Bank
Supervision," 90, *Banking Law Journal,* August 1973, pp. 649, 663.
30. See note 22 *supra.*
31. Letter to General W. T. Sherman, December 24, 1868, in T. Sherman,
The Sherman Letters, (1894), p. 325.
32. Quoted in Gerald T. Dunne, "A Christmas Present for the President,"
in Richard A. Ward, *Monetary Theory and Policy,* 1966, p. 44.

This is the high adventure of the new day . . . We shall create and not destroy. We shall deal with our economic system as it is and may be modified, not as it might be if we had a clean sheet to write upon, and step by step we shall make it what it should be in the spirit of those who doubt their own wisdom and seek counsel and knowledge, not shallow self-satisfaction or excitement of excursions whither they cannot tell.[33]

33. *Ibid.*, pp. 56–57.

APPENDIX

AN ACT TO DECLARE A NATIONAL POLICY AGAINST INFLATION AND FOR OTHER PURPOSES

Short Title

Section 1. This act may be cited as "The Anti-Inflation Act of 1975."

Declaration of Policy

Section 2. The Congress finds that inflation destroys employment opportunity, impoverishes old age for anyone on a fixed income, expropriates savings, and accordingly hereby declares that it is the continuing policy and responsibility of the Federal Government to use all practical means to end inflation by stabilizing the value of the dollar and assuring its purchasing power.

Employment Act of 1946

Section 3. The Employment Act of 1946 is hereby amended
   (a) by inserting the words "stabilized and assured" before "purchasing power" wherever the latter phrase appears in such Act;
   (b) by adding at the end of Section 3(a) thereof: "including his recommendation for the rates or range of rates for annual increases in the money supply (herein defined as demand deposits in banks and currency in the hands of the public)";
   (c) by adding thereto a new Section 6: "Within 30 days of submission of the Economic Report of the President, the Congress shall by joint resolution set the rate or range of annual rates by which the money supply of the United States shall be increased."

Federal Reserve Act

Section 4. Section (c) of Section 12A of the Federal Reserve Act is hereby repealed and the following language substituted:

The time, character, and volume of all open market operations shall be governed with a view to the provisions of the Employment Act of 1916, as revised.

Response to Comments

T. Nicolaus Tideman

Prof. Buchanan and I have divided the task of responding to the comments on our paper along these lines: I will respond to technical points and he will respond to the philosophical issues.

The one point that Professor Dunne takes issue with is whether the time is ripe for change. We persist in our view that revolution is not at hand, though we would be delighted to be proven wrong.

Turning to Professor Winter's comments, we would like to emphasize that the question of the dependence or independence of gold clauses from the underlying contracts is not a simple one. We would agree that pure bets on government action will create undesirable incentives to influence that action, and need not be protected. However, some bets on government action are actually hedges that offset otherwise unavoidable risks. The question of whether any particular type of bet on government action should be protected must depend on a weighing of these offsetting factors—the value of hedges versus the cost of marketable redistribution from government action. When inflation is the risk in a "gold standard" world, gold clauses have a strong claim, though under the collapsing, "managed gold standard" of the 1930s they were an anachronism. We would support Professor Winter's proposal for constitutional protection for index clauses, but we would be inclined to let individuals decide for themselves whether to contract with the government on a nominal or an indexed basis.

Professor Johnson says that we should not use a commodity money as a standard for comparison with fractional reserve systems because a commodity money is internally inconsistent and contains the seeds of its own destruction. Some clarification is warranted. It was not our purpose to condemn fractional reserve systems, but only to point out how they differ from the commodity systems that precede them.

Johnson says that we distract attention from the central broad constitutional issues by starting our story with the disastrous situation in 1933. We would argue that Roosevelt and his Administration, who took over in 1933, were responsible for the revolution in the monetary constitution and were not responsible for prior events, so that in appraising the changes it does make sense to start with the situation in 1933.

Professor Friedman says that the events between 1930 and 1933 did not demonstrate the susceptibility of a fractional reserve currency to losses of confidence, although he obviously accepts this with respect to deposits. We agree that the events of this period did not, empirically, demonstrate this susceptibility with respect to currency. But it seems reasonable to suggest that the collapse of fractional reserve deposits called attention to the potential vulnerability of the fractional reserve currency, and that this recognition may well have influenced policy choices.

Friedman disputes our statement that continued adherence to the international gold standard in 1933 might have prevented or substantially delayed internal monetary expansion. He cites the facts that there was little demand for gold prior to the suspicion that Roosevelt would devalue, and that the ratio of the gold stock to the quantity of money was unusually high. We stand by our statement and support it with the following arguments. First, it seems to us that if monetary authorities are asking themselves whether they can expand the currency issue without provoking a loss of confidence in currency, the ratio they would look to would be the ratio of the government gold stock to currency in the hands of the public and not the ratio of the total gold stock to the quantity of money. Secondly, we are not convinced that rapid expansion in currency in March 1933 would have been consistent with continued adherence to convertability between dollars and gold at $20.67 per ounce. We can well imagine such an event provoking a shift from dollars to gold even without George Warren's recommendation of devaluation. The psychology of monetary events cannot be ignored. To support this point we quote from Albert J. Hettinger, Jr., a partner of Lazard Frères and Company and a director of the National Bureau of Economic Research, whose questioning comment on the Friedman-Schwartz interpretation of the events of 1929–33 is printed at the end of Friedman's and Schwartz's *Monetary History of the U.S.* Hettinger was a businessman during the Depression and felt compelled from his background to place greater weight on psychological and political factors than Friedman and Schwartz had. He said:

> It has been burned in upon me that monetary policy, in the final analysis, acts on men whose conduct is not predictable; it neither operates in a vacuum nor in a world in which all other factors can be taken as constant.

Later he says,

> With respect to the final statement that the collapse of the monetary system was unnecessary, this I cannot feel has been proved. To me, each move in the high-powered-money arsenal involves a calculated risk. If its impact on men's minds is favorable, possibly even if it is neutral, the arithemetical results postulated by the authors follow as night after day. . . . If those moves were deemed inflationary and "unsound," the results could have been other than those desired. In that day a citizen fearing devaluation could choose gold rather than paper, and the international flow of gold, seeking safety, was as unpredictable as that of a gun loose on a battleship pitching in heavy seas. The authors may well be right; they are outstanding monetary economists—but I would prefer the terms "possibly" or conceivably "probably" rather than "clearly" need not have happened.

We share Hettinger's uncertainty. Furthermore, we would stress that decisions had to be made from the understanding of possibilities that the actors of that time possessed. A shrewd Jacob Viner may well have seen possibilities that others would not see.

Finally we would argue that the flood of gold that entered the U.S. at $35 per ounce reflected not only the artificiality of that price but also the growing political uncertainties of Europe. It cannot all be attributed to the excessiveness of the price in 1934.

We concede Friedman's point that the government did not succeed in capturing all of the profit from devaluation. We do not claim to justify, as Friedman seems to think we do, the expropriation of gold without compensation.

On the question of whether there is justification for prohibiting the circulation of gold in competition with currency, we would first repeat our belief that there must be some uncertainty regarding the psychological factors that would have operated in alternative scenarios. Secondly, we would dispute Friedman's statement that the loss of confidence in banks enhanced confidence in the national currency.

Although it is true that people fled from deposits to currency, we would interpret this shift in demand as arising from a decline in the value of a substitute for currency (namely deposits), rather than from heightened confidence in currency.

We acknowledge Friedman's point that the greater stability of competing currencies can reduce the costs of unstable national currencies. Nevertheless, we do not regard this factor as unambiguously outweighing the transacting costs of multiple currencies. We see this as an issue where people could reasonably favor either national monopoly or competition.

With respect to indexation, we do not think that there are such great differences between Friedman and us. We agree with Friedman that it would be desirable to protect index clauses, but we simply do not expect the courts to find that they must protect them.

Response to Comments

James M. Buchanan

I want to confine my response to two points. The first has to do with the "choice setting" ambiguity that was noted by Professor Winter, which may be interpreted as encompassing the point concerning predictability made by Professor Johnson. We acknowledge the ambiguity in our paper here; let us try to clarify our own position as best we can.

In Section V of our paper we discussed what we called the "contractual origins" of monetary rules or monetary institutions. We asked the question: What are the elements of a monetary order that might be predicted to emerge from a genuine constitutional or contractual setting where quasi-permanent rules are to be chosen and in which participants are uncertain as to their own positions? We argued that, minimally, two elements might be predicted to emerge here. The first would be that the government would be empowered to define the monetary unit. To us, there seem to be "public good" characteristics of money that would be recognized and embodied in such a rule. Secondly, we argued that if the government should be empowered to regulate the value of money its objective should be that of insuring predictability. Note that these are minimal characteristics, and that, within these limits, there may be many separate details of monetary rules. These elements, taken alone, do not tell us whether an automatic or a managed monetary system would emerge. I do not think we can derive a particular structure from this basic contractarian framework, and I do not want to make the same mistake that I think John Rawls made in trying to be too specific.

These are, in one sense, predicted characteristics of any set of monetary institutions that would emerge from a true contractual setting. So far, so good. But where we are ambiguous in our paper is in seeming to confuse this with the actual constitutional structure that might have been inferred as being in existence in the United States. The question that Professor Winter poses is the relevant one: Should a court or a judge evaluate proposed changes in monetary rules on the basis of an idealized constitutional contract independently of the actual historical evolution of law, for which the idealized construction is not at all descriptive? Professor Johnson's criticism suggests that we may have fudged on this question by implying that the idealized construction is, in fact, descriptive of the historical record in which case, of course, there is no question. But if and when we acknowledge the differences between the legal history and the idealization, do we want the courts to adopt our own idealization at the expense of the record?

Our answer is that we do not, and we erred if this interpretation seems to be an appropriate one to place on this part of our paper. I find myself in a position close to that which I put to John Rawls two years ago. In a review article on his theory of justice I asked: Would John Rawls want Earl Warren to implement the Rawlsian precepts of justice independently of any explicit collective decision embodying these precepts more formally?[1] To my satisfaction, Rawls responded to me that he would not do so. But, as Professor Winter points out, many persons do interpret John Rawls as implying that he would, in fact, want the Earl Warrens of the world to do just that.

To return specifically to monetary rules here, we find our treatment in the last part of Section V of our paper to be misleading in this respect. In one sentence, we use the phrase "duly constituted authority" along with the phrase "contractually-justifiable" as if these were the same thing. Obviously, they are not, and should not be treated as if they are. In other words, we should have been unhappy in the large, even if satisfied in the small, if Mr. Kenner's legal action should have succeeded. Properly interpreted, our argument in this section of the paper should be read as a plea for explicit movement toward the enactment of contractually-legitimate rules, enactment through appropriately constituted decision-making authorities, and not through the instruments of judicial legislation.

My second main comment is in response to Professor Milton Friedman's more fundamental criticism of our paper. I want here to confine my remarks to what seems a basic difference in approaching history. It seems to me that we must impose an order on historical events by a mental process that depends only in part on the empirical record that may be objectively measured. As this record becomes more complete we must, of course, modify our vision of historical events, but a vision it must remain. As one of my colleagues put it, we must always preselect a functional form for our regression. There is no unique solution to the historical game, and it seems

---

1. See my, "Rawls on Justice as Fairness," *Public Choice*, XIII (Fall 1972), pp. 123–128.

misleading to suggest otherwise. To be slightly more specific here, an understanding of the events of the 1930s requires us to try, as best we can, to understand what the men and women who were the actors on that stage believed, to make some attempt to discover their own vision of the economic and political world in which they lived and acted. Professor Friedman seems to us to imply that these men and women "should have" seen their world as it now appears to him, and that their apparent failure to do so proves their culpable stupidity or their malevolence.

We are more humble in our approach. The decision-makers in the 1930s were neither more nor less stupid or evil, on average, than those we have with us today. And there is simply too much that is subjective in the environment for even an idealized accumulation of the empirical record to overcome. Professor Friedman may well be correct in his "judgment" (and we emphasize that it remains a "judgment") that an omniscient and benevolent group of decision-makers, or even a Benjamin Strong, could have prevented the monetary debacle that occurred *without* suspending convertibility, *without* restricting the ownership of gold, and *without* abrogating gold clauses in contracts. (We all should surely defer to Professor Friedman's vastly superior expertise in reaching such judgments.) But whether or not such an idealized set of decision-makers could have done this is not, to us, the central question. Instead of this we ask: In the economic and political environment in which they found themselves, did the actual and effective decision-makers consider the course of action now seen to be most desirable by Professor Friedman (and perhaps by everyone of us here) to be a viable alternative? In our paper, we do not seek to "justify" the action that was taken. But we do seek to "understand" this action, within limits.

There are significantly different implications to be drawn from the approach taken by Professor Friedman and that which we take in the paper. In our framework, the reform emphasis is necessarily institutional, and discussion centers on the prospects for modifying the structure so as to prevent recurrence of a sequence of events that, in retrospect, we all adjudge to have been tragic. In this respect, we should defend our emphasis on the vulnerability of any monetary structure, domestic or international, that embodies fractional reserves. Whether there was or was not a danger to gold reserves in 1933 is not directly germaine to the basic issue. Given the institutions as they existed, the fear of the consequences of a run on gold may have been a basic component of the actions that were taken.

Finally, and hopefully, it is in this last sense that Professor Friedman seems to be somewhat inconsistent with himself. The Milton Friedman associated with the fixed monetary growth rule seems to us to be on our side in almost all of the differences herein discussed. This rule represents institutional reform, one that aims at replacing the discretion of fallible men with a predictable rule of law. In viewing the history of the 1930s, we are struck with the absence of such predictability, with the instability of the institutions in being. In his critique, Professor Friedman seems, by contrast, to be struck by the stupidity and malevolence of the decision-makers. We suggest that our approach is more relevant to the problems that we face today. We shall always have with us decision-makers who fall short of Professor Friedman's standards; but institutions may be reformed.

Conference
Participants

WILLIAM H. ADAMS, Attorney at Law, Mahoney, Hadlow, Chambers and Adams, Jacksonville, Florida

ARMEN ALCHIAN, Professor of Economics, University of California, Los Angeles

MARTIN ANDERSON, Senior Fellow, Hoover Institution on War, Revolution and Peace

MANUEL AYAU, President, Universidad Francisco Marroquin, Guatemala

LEMUEL BOULWARE, Delray Beach, Florida

KARL BRUNNER, Professor of Economics, Graduate School of Business, University of Rochester

JAMES BUCHANAN, University Professor and General Director, Center for the Study of Public Choice, Virginia Polytechnic Institute and State University

LESTER CHANDLER, Professor of Economics, Princeton University

DENNIS P. CLUM, Executive Vice President and First Officer,

The First State Bank of Miami, Miami, Florida

KENNETH W. DAM, Professor of Law, University of Chicago Law School

EDWARD DAUER, Visiting Associate Professor of Law, Yale Law School

HAROLD DEMSETZ, Professor of Economics, University of California

GOTTFRIED DIETZE, Professor of Political Science, Johns Hopkins University

GERALD P. DUNNE, Professor of Law, St. Louis University Law School

RICHARD EPSTEIN, Professor of Law, University of Chicago Law School

MILTON FRIEDMAN, Professor of Economics, University of Chicago

GOTTFRIED HARBERLER, Adjunct Scholar, American Enterprise Institute for Public Policy Research, Washington, D.C.

WESLEY H. HILLENDAHL, Vice President and Economist, Bank of Hawaii, Honolulu

HENRY MARK HOLZER, Associate Professor of Law, Brooklyn Law School

JOHN O. HONNOLD, JR., William A. Schnader Professor of Commercial Law, University of Pennsylvania Law School

HARRY G. JOHNSON, Charles F. Grey Distinguished Service Professor of Economics, University of Chicago

RONALD W. JONES, Munro Professor of Economics, University of Rochester

DAVIS E. KEELER, Institute for Humane Studies, Menlo Park, California

DONALD KEMMERER, Professor of Economics, University of Illinois, and President, Committee for Monetary Research and Education, Inc., Champaign, Illinois

ISRAEL M. KIRZNER, Professor of Economics, New York University

EDMUND W. KITCH, Professor of Law, University of Chicago Law School

ARTHUR A. LEFF, Professor of Law, Yale Law School

MICHAEL LEVINE, Professor of Law, University of Southern California, and Henry R. Luce Professor of Law and Social Change in the Technological Society, California Institute of Technology

HENRY G. MANNE, Distinguished Professor of Law and Director, Center for Studies in Law and Economics, University of Miami Law School

HAROLD L. MARQUIS, Professor of Law, Emory University Law School

JERRY L. MASHAW, Professor of Law, University of Virginia Law School

DAVID MEISELMAN, Professor of Economics, Virginia Polytechnic Institute and State University

ALLAN H. MELTZER, Maurice Falk Professor of Economics and Social Science, Graduate School of Industrial Administration, Carnegie-Mellon University

ROGER LEROY MILLER, Professor of Economics and Associate Director, Center for Studies in Law and Economics, University of Miami Law School

A. NEIL MCLEOD, Executive Director, Liberty Fund, Inc., Indianapolis, Indiana

SOIA MENTSCHIKOFF, Dean, University of Miami Law School

JAMES S. MOFSKY, Professor of Law, University of Miami Law School

WARREN NUTTER, Paul G. McIntire Professor of Economics, University of Virginia

ALAN REYNOLDS, Contributing Editor, *The National Review*

BEN A. ROGGE, Professor of Economics, Wabash College

RICHARD N. ROSETT, Dean, Graduate School of Business, University of Chicago

JOHN RYAN, Indianapolis, Indiana

TERRANCE SANDALOW, Professor of Law, University of Michigan Law School

HELEN E. SCHULTZ, President, Liberty Fund, Inc., Indianapolis, Indiana

WARREN F. SCHWARTZ, Professor of Law, University of Virginia Law School

ARTHUR SHENFIELD, Visiting Distinguished Professor of Economics, Rockford College

PHILIP SCHUCHMAN, Professor of Law, University of Connecticut Law School

HENRY KING STANFORD, President, University of Miami

NICOLAUS T. TIDEMAN, Visiting Associate Professor, Virginia Polytechnic Institute and State University

GORDON TULLOCK, Professor of Economics, Virginia Polytechnic Institute and State University

GEORGE VOLSKY, Coral Gables, Florida

JOHN WEISTART, Professor of Law, Duke University Law School

RABURN M. WILLIAMS, Visiting Associate Professor, Center for Studies in Law and Economics, University of Miami Law School

RALPH K. WINTER, JR., Professor of Economics, Yale Law School

RENE A. WORMSER, Attorney at Law, Wormser, Kiely, Alessandroni & McCann, New York, New York

CHARLES J. ZWICK, President, Southeast Banking Corporation, Miami, Florida

Discussion

I. The Panel's Comments

JOHNSON: I should like to raise the question of what economists during the 1930s believed about gold. I remember reading very carefully the accounts of the suspension of sterling convertibility. For some months after that, the market fully anticipated that all prices in the United Kingdom would rise in proportion to the fall in the value of the pound. That obviously was nonsense, given the depressed conditions, but it does, it seems to me, raise an important question— whether it was thought that raising the price of gold was going to do something by itself.

Now Friedman, at some point in his paper, suggests that if the theory had been right, the domestic price level would have risen proportionally to the devaluation of the dollar in terms of gold. I do not think that theory tells us that, given the government's recognition of mass unemployment and unemployed resources.

But it does seem to me crucial to know what Roosevelt thought he was doing, or whether he thought he was getting at anything by changing the price of gold. If he really felt that all prices in the U.S. would rise, then that would put one complexion on the effect of the abrogations. If he did not believe that, it is something else again. If he felt prices were going to rise with the rise of the price of gold, then he was deliberately injuring the individuals who had attempted to guard themselves against that.

If, on the other hand, he did not expect prices to rise, then letting private citizens have the benefit of the increase in the value of gold for private contracts would have meant allowing a deliberate, arbitrary redistribution of property. So I think it might be worthwhile to discuss what people at that time thought a rise in the price of gold would do.

TIDEMAN: My understanding is that Roosevelt did not specify in detail what he thought was going to happen, and that the price of $35 an ounce was not declared until nine months after the gold had been called in.

In the intervening time, there were governmental actions to try to raise the price of gold in the foreign exchange market. It seems to me that what they were trying to do most directly was to raise domestic agricultural prices by lowering the foreign cost of dollars. I understand that George Warren had said that domestic prices would rise as a result. This is what standard economic theory would say as well, if the quantity of money increased in proportion to the rise in the price of gold.

However, despite the fall in prices, there was a strong anti-inflation sentiment in Roosevelt and others in the government, though not in a majority in Congress. There was some fear of excessive inflation, perhaps arising out of the German experience, which was not all that old. So they were trying to steer a course that would avoid a loss of confidence in money by doing things that could be called reckless, while ensuring that agricultural prices would rise. I think they were not really concerned with the impact of gold actions on domestic prices.

FRIEDMAN: I think that so far as George Warren was concerned, and he was the one whose policy was really being carried out by Roosevelt, there is no doubt about what Tideman has said. Warren was very explicit in saying that he believed in the monetary theory of the balance of payments. He believed that there was one world price for world-traded commodities, that the price of wheat was a world price, and that the U.S. was only a small part of the total world market. Therefore, in connection with gold his analysis would have been correct in a gold-standard world in which the U.S. held a small part of the total stock and the U.S. alone had raised the price of gold. Then the price of internationally traded commodities would immediately have jumped in dollars by the amount of the rising price of gold.

That did not happen because there were many other countries that were not on the gold standard. You did not have a gold-standard world. Great Britain was not on gold, while France was, and in addition, the U.S. did not hold a small part of total world supply; it was a large part. Still some portion of Warren's prediction did happen: there was an immediate response in U.S. prices of world-traded commodities. If you look at wholesale price indexes at that period, they behaved very differently from cost-of-living index numbers; and there is no doubt that there was a sharp change in the ratio of world traded to non-world traded prices as a result of the rise in the price of gold.

Now there is no doubt that Roosevelt wanted to raise prices. He was very explicit about that. I do not believe there were significant strands in the government discussions at that time of fears of inflation. They were mostly in the banking community, or they came from France; but they were not in Washington. Congress wanted Roosevelt to print some currency, but he would not print the currency that they authorized him to print in the Thomas Amendment of the Agricultural Adjustment Act. He was not a quantity theorist in that sense, and neither was Warren. Warren was not very explicit about this happening via the quantity of money. He was a "monetary-theory-of-the-balance-of-payments" man, who saw reflation happening via the change in the world price structure, rather than through gold.

Going back to Buchanan's comments [See "Response to Comments" Chapter.] many economists at the time, including Jacob Viner, believed that it was perfectly possible to inflate while staying on gold. But I agree with Buchanan that "conjectural history" in the sense he described it is the only kind there is. When I was speaking of his history being "conjectural," I meant something else. I meant that the historical statement did not correspond to historical fact, and I think both of us would agree that the conjectural history about what might have happened has to be based on a correct appreciation of what did happen.

I also agree completely with him, of course, that I am in favor of a fixed monetary growth rule; that I believe in institutions, not men. As we stated in *The Monetary History,* although these particular men had produced this result at this time, given the system some other men would have done it at another time.

Also I think we are in fundamental agreement about the constitutional issue—that what we want is predictability and a rule that will give it. But I think we must support that on the basis of as valid an interpretation of the historical record of that period as we can, and therefore—going to where I disagreed, not where I agreed—while it is true that any system that embodies fractional reserves is vulnerable, there was no threat to the gold standard. I think Nic Tideman's argument that the central bankers should have looked at the gold/currency ratio rather than the gold/money ratio is false. They were concerned with the total quantity of money, and if you go back to the record at that time, they had a gold reserve requirement against deposits as well as a gold reserve requirement against notes. True, the note requirement was slightly higher but not very different. They were right in looking at gold against total money supply, not just against currency, and, obviously, there was always a possibility that there would be a flight from gold. France had threatened to withdraw gold several times in the preceding years, so I am not saying there were not those possibilities. But that was not the climate of opinion at that time.

If you go back to the end of 1932, before Roosevelt came in, gold was not the problem. You will not see any discussions about a flight from gold. That really takes place in January and February, 1933, after the rumor spread that F. D. R. was going to devalue gold because of Warren's influence on him.

So I think that it is fair to say that the problem was banks, not gold, and that the problem was domestic, not worldwide. The U.S. was the only country in the world that had large-scale runs on banks and failures of banks, and the shifts out of deposits into currency. The rest of the world was simply dragged down along with it. I think Buchanan also underestimates the extent to which our own Federal Reserve economists at the time were aware of what was going on. You had the Royal Bank of Canada in 1930, 1931, and 1932, in monthly reports, month and after month, pointing out exactly what was happening in the United States; you could not improve on it now. Many people at the time were aware of what was happening, and the Federal Reserve Bank in New York at the time was fully aware. There is no doubt about Carl Snyder, or about George L. Harrison, or about the whole group of the people there. They were fully aware that the real problem was the problem of expanding high-powered money in such a way as to avoid the runs on banks.

FRIEDMAN: On the historical side, it's not irrelevant that in 1929 to 1933 there is no doubt that if Congress had controlled monetary policy you would have had a better policy than we did. It was only under the pressure of Congress that the Federal Reserve Board, in 1932, finally undertook a large-scale bond-buying program.

One of the amusing elements in the record is what George Harrison, the President of the Federal Reserve Bank of New York, did at the meeting of the Open Market Committee in which they decided finally to do this. They were meeting on the last day of the statement week, and he asked them to make a special rule that they should call up the New York Bank and tell them to buy the bonds right away because he was going to testify in Congress the next day, and he wanted to be able to say to Congress: "We have already launched this program; we're not doing it under your pressure." They called up to New York and got them to buy a couple of hundred million dollars of bonds so that he could report that they were launching the program. And they ended the program, two weeks after Congress adjourned, when the pressure was off.

WINTER: I'd like to ask Professor Dunne a question. He referred in his comments to the fact that this was the only Supreme Court decision that President Roosevelt was prepared to override. I'd like him to elaborate on that.

DUNNE: I think we see the evidence in memoirs of the old New Dealers that are surfacing now. For instance, F. D. R. had issued instructions, I understand, to Ambassador [Joseph P.] Kennedy, who was then Chairman of the SEC, and to the Reconstruction Finance Commission, that they were to proceed as if the government was upheld before the Supreme Court.

WINTER: I think that's incorrect. I do not think he was prepared to disobey a Court order in a particular case, that is to say, where a particular plaintiff has sued the government and has a final court order directing the government to pay. I don't think he was prepared to do that. I believe he was prepared to find every way he could to go ahead on the old rules, where there was not a direct court order, and also to try to get the Court to overrule itself or to take other legal action. But I think this an important distinction from what Dunne suggested.

DUNNE: Well, I didn't mean to say that he had instructed the Treasury of the United States not to pay a judgment, if indeed one had been handed down.

WINTER: But that's a big difference. It is essentially the difference in the position President Lincoln took toward the Dred-Scott decision, as against the position Lincoln took on a couple of *habeas corpus* cases where he directly disobeyed a court order.

MANNE: May I ask Ralph Winter a question about this gold-clause business? Do I understand you correctly to be saying that independently, of any monetary clause in the Constitution, there would still have been no question about the monetary authority of the government. You think that a court might have abrogated a gold-clause agreement between two private parties on the ground that the results were not those contemplated at the time the agreement was entered into?

WINTER: Yes, I believe my conclusion is that a common law court should have held that the doctrine of frustration applied, and that the gold clauses ought not be enforced. On the other hand, some writers took the view that the gold-clause abrogation by the court was premised on gold as money, not on a failure of expectations.

DUNNE: But there is a passage in Chief Justice Hughes' opinion where he says, as I remember it, that the government can prevent trading in opium, as it had, and had frustrated expectations there. He concluded that this is what was done here, and he draws the analogy between prohibition of the drug traffic and prohibition of the gold traffic.

MANNE: I'd like to raise the point for Ralph Winter. My instincts as a lawyer tell me that the doctrine of frustration may have a qualification in a case in which a party in interest is, in fact, the one who has created the frustration of purpose, and that's exactly what existed in this situation, certainly with the government's own obligations and in a sense even with the private contracts, since even there we might want to "police" the government with a rule that might supercede the doctrine of frustration of contract.

WINTER: I'm not sure what you mean. Professor Friedman has raised the question, what happens if Congress had not passed the resolution abrogating the gold clauses and suits were then brought in the forty-eight different states on the gold clauses. I responded that in that situation a common law court might well have applied the doctrine of frustration of contracts. I can't see what kind of policing it would be for the Florida Supreme Court to say, well, frustration doesn't apply; we're going to enforce the clauses. For what purpose, Henry [Manne]? How does that police the federal government?

MANNE: Simply by requesting the government's interference with private hedging or betting agreements, by creating conditions that would then let a frustration defense succeed. It's not clear to me that these contracts should not be enforced.

JOHNSON: It seems to me that we've reached a point where we really need a lot more historical research as to how these clauses originated. Often in legal contracts there is a perpetuation of language well past the phase in which it was originally relevant. For instance, one can see that a gold clause would have been a useful clause at the time of the suspension of 1860–79. Again, the U.S. really only went on the gold standard, officially, in 1900, so there some particular legal phrase might have been introduced saying now that gold is a standard in the U.S., we will make the contract in terms of that standard. Then such a clause is perpetuated for no other reason than for the sake of having the language there. The interpretation suggesting that people inserted gold clauses because of fear of inflation is something that doesn't seem to me to be particularly plausible historically.

FRIEDMAN: We know what the facts were about that, Harry. From 1860 to 1879, there were no gold clauses, but gold clauses were put into every bond issue thereafter. They started in 1879 as a result of the experience during the greenback episode, when gold sold freely in terms of greenbacks at a free-market price; the gold clauses said that you were to receive the number of dollars required to buy X ounces of gold of such and such weight and fineness. It was in contemplation of a return to such a situation—one in which the U.S. would be driven off gold and you would have a free greenback price of gold—that gold clauses were introduced in bonds after 1879. Almost every bond issued by both the private sector and the government from 1879 on had a gold clause, and it's perfectly clear that that was why they did it. The Liberty Bonds in World War I had a gold clause in them and were sold on the basis of the gold-clause protection.

MANNE: Ralph, given this economic history, it seems to me that my argument for enforcing these clauses is stronger than I had thought. It is, after all, hard to believe that the presence of those clauses was simply "unnoticed" by all the Wall Street bond lawyers.

WINTER: It seems to me to give more reason to uphold the abrogation, or to apply the doctrine of frustration. It is curious that Chief Justice Hughes' opinion in a private case (the *Norman* case) begins with the question, what do these clauses mean? So even though they had been around for years, they were still that ambiguous.

But he did conclude, as I recall, that what they meant was that someone was entitled to however much currency it required to buy the stated amount of gold. That is, it was a clause worrying about a depreciated currency, not a clause saying that I'm only interested in holding gold. That is, the clauses had to be read as being designed to protect against changes in the value of money, and, therefore, I think, again, that the doctrine of frustration might reasonably apply. Further, it seems reasonable that Congress, having acted the way it did, having bought up gold and taken it off the market, might reasonably have feared that not all the common law courts would get the message. Therefore, as an adjunct to their action, Congress had to protect everybody, and so it abrogated all of the clauses.

MANNE: But you seem to forget that it was the Court, not Congress, that supplied your hypothesized corrective and thus would seem to have arrogated to itself the function of legislating in this area.

WINTER: No, it just upheld what Congress did.

MANNE: . . . I think my point is quite correct that a series of common law courts would not, uniformly, have reached the same result.

TIDEMAN: Can I go back to the issue Friedman talked about in his comments with respect to whether there was confiscation by the government and, if so, whether there was any justification for it. The gold was called in early on when price was $20.67, or perhaps a little higher in some foreign markets. Now suppose that the government had said, okay, you have a claim on us, we'll hold your gold and then we'll give you your claim at some later time when we decide what the new price is. And then suppose that the U.S. had reestablished convertibility of gold at $15 an ounce, and the government had said, now we're only going to pay you $15 an ounce, because that's all it's worth. There's at least some logic to saying that if the government paid what the article was worth at the time they called it in, then any later market price, higher or lower, should be appropriate.

FRIEDMAN: In the first place, its price at the time it was called in was not $20.67, it was something above that. Moreover, it was clear that the intent of the government was to raise the price still higher; it was clear that the government called it in for the sole purpose of preventing private people from profiting from the rise in the price of gold that it was intending to bring about.

FRIEDMAN: But, Nic [Tideman], let me ask you a different question. At the same time the government was involved in raising the price of silver, but there was no calling in of silver. Private people who held silver in any form whatsoever benefited from the rise in the price of silver. There was no talk of calling in silver. They did the same thing with wheat.

Now what was the difference between what they did with wheat and what they did with gold?

WINTER: I think the constitutional argument is made as follows: It may be that everything Milton Friedman has argued about expediency and the lack of justifications, etc., may be absolutely correct, and I found his argument persuasive. But I'm not sure that a court ought to review the expediency of a decision, and I would take the position that with its power to set the value of money and with gold having a monetary function, Congress has the power in question, and it may use it for stupid reasons, for no reasons or for bad reasons. But *that* is not for the Court to review. That doesn't mean that there may not have been a confiscation in the sense that Professor Friedman talked about it. It's just that Congress's power to confiscate may be substantial.

JOHNSON: Okay, let me see if I can give an answer to Friedman's point. I think that I would prefer to have a government that had no power to manipulate prices, but I think that we didn't have that, and I can see a possible rationale for people feeling that if government is going to manipulate prices, the government should stand to get all the gain. Now the rationale that I see for this is that if the government is both manipulating prices and letting other people try to outthink the government or profit from the government's attempt at manipulation, (1) it will take more tax dollars to produce a given amount of manipulation, and (2) there will be very strong financial gains for leaks about the intentions of the government. And so if you get to the undesirable position where governments are going to be manipulating prices, it may be less unattractive to have the government have coercive powers to keep people from interfering with its efforts to engage in that activity.

FRIEDMAN: It seems now that the Buchanan-Tideman and the Ralph Winter positions are different. If I understand Ralph Winter, he is saying that this is just a special case of the power to tax, and he is saying that Congress, in effect, was exercising its power of imposing a tax of $35 minus $20.67 on all holdings of gold.

WINTER: No, I have been saying that Congress can change the value of the money people are holding, and as long as gold has a monetary function, that is one of the things it can change the value of, or, indeed, eliminate its value, as it did with some rationing stamps in World War II.

MANNE: Ralph [Winter], is your *ideal* constitution as open-ended as your view of the one we have?

WINTER: Well, I am not sure. There are constraints that we can place in an ideal constitution that could prevent what we are talking about. I am rather convinced that we probably ought to protect indexing, but would protecting indexing alone accomplish everything we want?

TIDEMAN: Let me ask a different question. Suppose that Congress had passed a law requiring people to turn over silver, or let me not take silver, take something that had no monetary connection whatsoever—wheat. Would that have been upheld by the Supreme Court at that time?

WINTER: Probably not at that time. Within three years it would have been. It was, but not right at that time and principally, I think, because wheat did not involve the monetary power that gold did.

TIDEMAN: Well, that is all right, but then I think we are not in agreement. We are saying that if you rule out the monetary excuse for doing it, then it was straight confiscation, and the constitution should have prevented it, while you see a very different proposition. You are making the proposition that you should not forbid it under those circumstances.

WINTER: Yes, that is why I would say that if I were going to give the government power to manipulate prices (which I would not do), I would also give it the power to confiscate any individual holdings at the prevailing prices at the time it starts this manipulation.

FRIEDMAN: Perhaps we should think along the lines of what Jim Buchanan was saying earlier about looking at institutional effects and looking at the consequences from that point of view. The prohibition of private ownership of gold in 1933 meant that the gold standard was through. It was only a question of time, then, of when you got to a fiduciary standard.

MANNE: And the constitutional implication that follows from that is that the court should have treated it as wheat, or lead, or any other commodity. But that lies very much at the heart of the various arguments that have been made about the loss of the public good, the benefit of government-regulated currency, and the government's obligation for protecting the stability of values of currency.

FRIEDMAN: Well, I think the real, fundamental constitutional change was made in 1914, and that, in a way, what happened in 1933 or '34 was an inevitable consequence of that. This is why I disagree with Gerry Dunne and think that the U.S. would have been far better off if the Federal Reserve System had never been established. It seems to me that the Federal Reserve fundamentally changed the whole system. We left anything like a real gold standard, anything even approaching it, and we shifted onto a managed standard, which was predominantly a fiduciary standard, and in which gold was playing a very subsidiary role. That this was true in the 1920s when the Federal Reserve System sterilized the gold inflow, and it was true in the 1930s when again they sterilized gold and did not react in response to it. We would have been better off in both the '20s and '30s as a whole if we had gone to either of the extremes of continuing with a pre-1914 real gold standard or shifted in 1914 to a pure fiduciary standard.

MANNE: I wonder if there is a third alternative to what Milton Friedman has offered that might even be serving us better today—that is, if the Supreme Court decided those *Gold -Clause Cases* differently. It seems to me that there were policy implications that have not been fully explored here in the broader terms that Milton is now discussing. I realize that that is not the traditional outlook of courts and particularly the courts of that era, but it still seems to me it might have been preferable.

BUCHANAN: It seems to me, though, that one cannot possibly read that record without coming out in support of the majority opinion. The minority presented a very weak case. I want to go back to one point that Friedman made in his comment and also in his paper. It relates to our somewhat divergent reading of the history. I think we have come a good deal closer together in this discussion, but both in the paper and in one of his comments he said that there need not have been this gold problem, and we agree fundamentally here, but once Roosevelt was elected and once it was known that he was talking to Warren, then what we say historically in the paper seems to be upheld.

FRIEDMAN: No, not at all. If Roosevelt had not adopted the policy he did, even after all these rumors things would have been different. Let's say in March, 1933, if he had said, "We are going to restore the convertibility of gold immediately," whether at $20.67 or at $35, he could have carried that through. There would have been no run if he had done that. It was because everybody knew he was not going to do that which caused the problem.

DUNNE: You may recall, in the closing days of February, President Hoover asked him for an explicit statement, which he refused to give, so you can say Herbert Hoover precipitated the whole situation by asking the wrong question.

MANNE: I think Friedman also pushed it back to J. P. Morgan.

FRIEDMAN: Well, that is a different question; it concerns one of the most fascinating little bits of history—it is not conjectural, it is true—about the failure of the Bank of the United States on December 11, 1930. It failed because J. Pierpont Morgan's father had been cross-examined in the Peugot investigation by Samuel Untermeyer and had died three months later and that's why the Bank of the United States failed on December 11, 1930.

MANNE: Do you think that that failure was instrumental in creating the fears that generated the problem we have been discussing?

FRIEDMAN: Oh, there is no doubt. It was the failure of the Bank of the United States on December 11, 1930, that produced a series of bank runs that, in turn, really produced the flight out of the banks. Now maybe another bank would have failed later on. I'm not saying one way or the other. I really do not want to make too much of it because I think it was this particular failure that at that moment triggered the first bad series of bank runs. But if the Bank of the

United States had not fallen at that time, it would have
fallen later on.

II. The Role of Gold and Economic History
Underlying the Gold Cases

MELTZER: I wanted to raise the question about the role of
gold in the Federal Reserve's reasoning. Our reading of
the records of the Board of Governors of the Federal Reserve
led us to conclude on this point that the interpretation of
Friedman and Schwartz in their *Monetary History of the United
States* was incorrect, that in fact gold played a larger role
in the behavior of the Federal Reserve System prior to
1933–34 than the *Monetary History* would lead one to believe.
Several episodes point in that direction. On the broadest
issue of the role of free gold in the Federal Reserve's policy
determination, I would like to defend Buchanan and Tideman on that point. Karl Brunner and I have read the detailed
minutes of the Board, and the tape that was received from
Milton Friedman and Anna Schwartz. A reading of this
information leads us to conclude that on this point Friedman
and Schwartz are wrong. They maintain that gold was not
a constraint on the Fed's ability to expand the money supply.

FRIEDMAN: I think it is important to separate two questions:
the role of gold from the point of view of the Federal Reserve's
free gold and the role of gold from the point of view of
the public's demand for gold. Those are two quite different
issues. What was the public's desire: to convert currency
into gold?

MELTZER: There is no evidence that the public wanted to,
as a matter of fact, do so on a large scale. But if the public
would have demanded gold, then there would have had
to have been a suspension of convertibility.

FRIEDMAN: On that we are agreed. On the other point, I
have nothing to add to the discussion of the *Monetary History;*
I think the evidence is very clear that free gold was an
excuse and not a reason for the Federal Reserve's behavior
and there is no doubt that McDougal was jealous in New
York; he wanted to protect his gold reserves in Chicago.
So far as the bond purchase operation of 1932 is concerned,
I think the evidence is very, very clear. That was instituted
under the pressure of Congress with the Glass-Steagall Act
as an excuse. The bond purchases were ended two weeks
after Congress adjourned, and they were ended because
the Fed never wanted to engage in them in the first place.
The Federal Reserve never wanted to go into that open
market operation. They went into it under Congressional
pressure and as soon as the Congressional pressure was over,
they stopped.

LEVINE: For the benefit of the illiterates in the audience,
what is the free gold problem?

FRIEDMAN: The free gold problem that I mentioned can be
understood by looking at the gold reserve requirement for
the Fed. It had to hold gold equal to a certain percentage
of its deposits. In addition to the 40 percent gold reserves
for currency, 60 percent of collateral was required, satisfied
either by gold or eligible paper. Not all assets qualified.
Only eligible paper qualified. In particular, government
bonds did not qualify as reserve backing, only gold or eligible
paper. So the Fed interpreted some of the gold in excess
of the 40 percent requirement as being tied down as further
reserve backing.
    Now the technical argument is over whether the Fed could
have gotten around this requirement. Our argument was
that there was always plenty of eligible paper available. Hence,
if the Fed had wanted to pull it in, they could have pulled

it in. For example, there were bankers acceptances outstanding; there was a market for bankers acceptances, and the
Fed could have raised a little bit the price they were offering
for bankers acceptances. There was never any trouble, in
my opinion, about having enough eligible paper. They held
down the eligible paper in the Fed's possession partly because
they wanted to tie their own hands. That is our interpretation,
so that is why I wanted to disagree.

CHANDLER: In talking about the role of gold, I would like
to put it in a broader content. If you all believe in democratic
processes, then you have to raise the questions: What are
the objectives that people will at least accept, if not most
people, a great many? And, secondly, what is the understanding in the period about the relationship between possible
means and possible ends? And, it seems to me, one has
to go back (and here I am going to follow Harry Johnson
fairly closely) to what happened, say, between 1900 and 1933.
There were several concepts of the value of money that
people could not make up their minds among. One was
the value of money in terms of gold, and they thought they
had settled that in the Gold Standard Act of 1900. But,
this obviously was not enough. Another meaning was the
value of the dollar in terms of foreign currency.

Now, they didn't know in that period what the relationships
among these would be. Some hoped that by fixing the value
of the dollar in terms of gold, you get the others automatically.

Then there is another concept of money value in a very
different sense of the term, namely the interest rate—the
price for the use of credit, in effect. This notion got incorpo-
rated into the Federal Reserve Act so you had all these
different concepts constantly vying with each other.

Now, I would claim that when the moment the Federal
Reserve was put in, and here I'm following Harry Johnson,
and I think also Milton Friedman, you immediately took

a move toward a managed money supply and away from an automatic gold standard. Well, through a whole series of events, running through the first half of the 1920s, the gold standard that we were on became both (a) non-international and (b) managed with deliberate efforts taken by Benjamin Strong and others to prevent gold movements from affecting the domestic money supply and presumably also, therefore, the price level domestically.

A great deal of Federal Reserve efforts during this period were devoted to another meaning of the term "value of the dollar," namely to re-establishing a system of exchange rates that would remain stable through time, which Friedman would say would be the next mistake that they made. But at the time it was (except possibly in England) not considered to be a mistake.

Throughout this whole period, they get themselves concerned with interest rates, which may not lead to a stable value of money.

Now, this is a little long-winded, I'm afraid, but I think it is a necessary background for the situation at the beginning of 1933, before Roosevelt came into power. There was only one thing that was very clear at that time, and that was that most people were burned by the increase in the purchasing power of the dollar. It is hard to know what index to use on that one, but by almost any index, the purchasing power of every dollar went up by 33-1/3 percent and by some others by as much as 50 percent. So, what do you have at the time Roosevelt comes into office? You have a banking system flat on its back, you have a purchasing power of the dollar at least 30 percent, and probably 40 percent for many purposes, above what it was in 1929. You have in the Congress hoppers filled with bills—saying, in effect, that we want something different: Some of them wanted greenbacks, some wanted silver, some wanted an increase in the price of gold, others wanted script, others wanted the Fed to buy three billion dollars' worth of governments, and this kind of thing. As a matter of fact, as most of you know, the Thomas Amendment that was put in to permit Mr. Roosevelt to do what he wanted to do, was really a defensive action on his part to escape being forced to do what he did not want to do—namely, adopt virtually all of these things. And yet there was still a strong gold standard flavor among people.

Now, with respect to the various things that were done at this point, I wish, along with Buchanan and Tideman, that Mr. Roosevelt had said to heck with gold, we are going straight on an inconvertible paper standard—in which case we would be talking about something else here today. But it was not possible at that time to do that, because he could not get the kind of backing he wanted. Possibly he could have with the landslide that he had, but there was enough talk within the next few years about "baloney" money and "phony" money and all that sort of thing that he wanted to keep the, if you will, aura of gold somehow or other to back up this type of thing.

The next question is, if he is going to call, why did he call in the gold? Suppose he wasn't worried about the outflows of gold. Suppose he was just worried about getting complete freedom—after all, this was a chance to get it. Well, get the gold in there and then there won't get to be a question, as he would have reasoned, about who gets the profits if anything happens to the price.

The next point I want to make is, I think—God only knows what George Warren's theory of money might have been—but it seems to me that the internal evidence indicates that one of the real things that Roosevelt wanted, and all the men around him wanted, was freedom to lower the exchange rate on the dollar relative not only to the pound sterling but to most of the other world currencies, except France and a couple of others. He wanted to get that exchange rate down.

Another thing that should be brought in here is the London Economic Conference of June 1933, which was a fiasco if there ever was one. The participants were talking about going back to some sort of a stable exchange rate system, and there was even a rumor that the U.S. and Britain were going to stabilize, on a temporary basis, at the pound sterling equal to $4. The furor that came from the Committee for the Nation and some of the others was tremendous. They said, "You've got to depreciate the dollar in terms of these foreign currencies a lot more than that."

Then came the gold purchase plan by the RFC [Reconstruction Finance Corporation]. As soon as it went into the international market, the purpose was clearly to depreciate the dollar in the market. Toward the end of 1933, when they had driven the pound up to about $5, the boys in Washington asked Harrison in New York to get in touch with Monty Norman at the Bank of England and say, "How would you like to stabilize here?" And Norman talked to the Chancellor of the Exchequer and said he had better not repeat too much of the conversation, with the dollar depreciated that much. So this was one of the major things about it. They loused up the use of the gold policy for the purpose of internal monetary expansion—they did not get that; but they did depreciate the dollar in the exchange market.

Now, with respect to continuing to hold the gold, after the devaluation of the dollar the price went up. One of the reasons this was done—I don't know all the reasons—was to keep freedom of action on the part of Roosevelt and his people to devalue the dollar again if necessary. They did not want to be in a position where they fixed the gold value of the dollar only to have some of the other countries decrease their currencies in terms of gold.

I further want to talk about the abrogation of the gold clauses. Just remember what had happened here. The price of gold went up almost 70 percent and also the price level had fallen in such a way that the dollar would buy about 40 percent more than before. You multiply 1.7 by 1.4 and yet get about 2.38. In other words, the real purchasing power that would have had to be paid by debtors to the creditors would have been up about 2.4 times as much as the creditors gave up if they lent in the late 1920s; farmers, of course, would have been hit even harder, because their indexes were down.

Well, I can't contemplate a Supreme Court in good conscience really saying, "You poor devils, go ahead and pay them 2 to 2-1/2 times what you got originally."

Furthermore, one of the great failures of the New Deal was to do anything to help business rehabilitate its financial position. They helped farmers, they helped homeowners, but business had to go along. And business after business was already bankrupt using any decent accounting system; this would have made it even worse and thrown still another obstacle in the way of recovery.

Just one final thing, I think always you've got to remember concerning this, what were the objectives? They were confused at the time about what their objectives were. What were their means of achieving objectives? They were equally confused on that. And I think when we talk about the monetary constitution for the future, we are again going to have to go into the question of what are the people's objectives, what kind of an array of value systems do they have, and what is the degree of understanding.

I would quarrel with Friedman, on one implication he left this morning, although I do not think he really meant to, and that was—Milton, this is probably not fair to you—but that people knew what *should* have been done. There certainly were people who knew. Most were guys in the good Yale tradition of Irving Fisher who would have come forth with good expansionary policies. But I ask you to go back to that period and look at the Economists' National Committee on Monetary Policy. Look at all the real-bills-doctrine people. They all said, in effect, don't do anything expansionary, or simply go to a real-bills kind of thing. And if you had had a poll of the economists of that time, in, say, 1932, Milton, I don't know how it would come out. There would certainly have been expansionists among them, but there were an awfully lot of guys talking in the other direction, too, and not just in the Fed.

KEMMERER: I will differ from a number of you. In the first place, let me answer a question about the Economists' National Committee on Monetary Policy. I was the last President of it, 1967–70. A group of economists who specialized in money, banking and international finance founded it in 1933, not in 1932 or 1931 as was alleged. They took it upon themselves to offer their expert advice to a government which they believed badly needed it. The group included Professors Ray Westerfield of Yale, O. W. M. Sprague of Harvard, my father, E. W. Kemmerer of Princeton, Walter E. Spahr of New York University, Neil Carothers of Lehigh, Col. Edward Harwood of the American Institute of Economic Research and several others, all well known economists at that time.

Yes, I believe they all would have agreed heartily that they did not like a rapid increase in the money supply at the time; in fact they were considerably worried over the haste with which the Congress or administration did a number of things. One action in particular, the year before, in 1932, was the passage of the first Glass-Steagall Act which authorized the substitution, temporarily, of government bonds for commercial paper as a secondary reserve against Federal

Reserve Notes or member bank legal reserves. This act is an important step in the monetization of our debt, which has been the biggest cause of inflation in the United States since that time. Temporarily there was a reason for the act but it was a major error to make it permanent. Members of the Economists' National Committee on Monetary Policy recognized that and argued against the law's permanency.

And I'd like to mention another development of the 1920s that we have since largely forgotten. There was a succession of organizations, all brainchildren of Professor Irving Fisher of Yale, that were desirous of finding ways to make the dollar stable in value. I have had occasion to look into this movement just recently. The first of these organizations was called "The Stable Money League." Fisher launched it in his own typical fashion. Having written a book on the subject, he had ex-President Frank A. Vanderlip of the National City Bank of New York give him a widely publicized banquet at which he spoke. People then flocked to him urging him to start an organization to do something about the instability of the dollar. He was more than willing.

His idea was to enact into law some mechanism that would keep the purchasing power of the dollar at an even level. The organization changed its name after about two years to the National Monetary Association because some members felt the original name simply made them disciples of Fisher. Then that organization broke down because of a dispute over what legislation to introduce. A third organization, the Stable Money Association, came on the scene devoted only to research and education; that in the early 1930s was replaced by the Committee for the Nation. But all four organizations had stable money as their goal and Irving Fisher was always in the background promoting them.

You have to put yourself back in that period to appreciate some of the thinking that was going on: it may be different from what you might suppose. Many of us today think of the '20s as a period in which the dollar was extraordinarily stable, and its purchasing power changed very little from 1922 to 1930. Nevertheless, many economists and others living at that time were worried over what the general price level might do next. They looked back over their life spans— remember they had been born in the 1860s, 1870s and 1880s—and they reflected on the roller-coaster ride, price-wise, that they had experienced. From 1865 to 1896 wholesale prices had fallen by two-thirds; from 1896 to 1914 they had risen by 60 percent; from 1914 to 1920 they more than doubled, and then in 1920–21 there was a 40 percent drop in just one year's time. Irving Fisher complained early in 1920 that the dollar had lost 72 percent of its buying power in just 24 years.

So the National Monetary Association had bills introduced into Congress by Alan Goldsborough of Maryland and later by James G. Strong of Kansas to do something to make the dollar stable. Congress quickly turned them down.

I am trying to emphasize that what George Warren and others were advocating, with some legislative success, was right along the same line. They went about it somewhat differently, but the stable purchasing power idea was very much in the air at the time.

Let me turn now directly to gold clauses themselves. The gold clauses that we talked about began as contracts payable in gold in 1868, the year of *Bronson v. Rodes,* as Friedman pointed out, and became "gold clauses" from about 1879 on, after specie payments had been resumed. They became more important in the 1890s when the very real danger arose that the United States would adopt bimetallism—that is what presidential candidate William Jennings Bryan advocated—that would in reality put the nation on a silver standard which could, in turn, result in considerable inflation. In short, the gold clause, as long as the nation was on a gold standard, was not particularly needed. Lenders and their lawyers were not worried as long as prices were stable, or even falling, and the country was on gold. They were concerned over the possibility that the nation would go off the gold standard, onto a silver standard or something else. And only if the nation went off the gold standard would these people expect to use the gold clause. Now what should be done if the problem became one of deflation—and it was that, 1865–1896—rather than inflation was not a problem the creditors considered. That after all was the debtors' problem. I think Buchanan is correct in pointing that out.

Let me add one last miscellaneous point and then I'll stop. In the 1920s the amount of gold coins in circulation, including coins in bank vaults, Christmas stockings and whatever, was a trifle over $100 millions, $115 millions in 1927. True, there were about $1 billion of gold certificates (backed 100% by gold in the form of ingots), so it did not take a great amount in the form of gold coins to make the gold coin standard work.

KEMMERER: The statement was made this morning that the gold flowing into the United States in the 1930s, especially after 1935, came largely from Europe, or at least very substantially from Europe, because of all the political uncertainties over there. Certainly a considerable amount did come for that reason but it is easy to exaggerate that aspect of the situation. We had $4 billions in dollars of the large (23.22 grains) size when we devalued and these became $6.8 billions of the smaller size (13.71 grains). You can, theoretically, mint 35 of these out of very 480 grain troy ounce. That is the basic reason for the $35 an ounce "price." Since the "price" of gold jumped sharply, it stimulated great activity on the part of companies mining gold. Their profits soared.

By 1940 American gold reserves of $6.8 billions in 1934 had grown to just $22 billions. The Western world's gold reserves had grown from $20 billions to $29 billions. In short, $6 billions of our new gold had come from Europe, $9 billions from gold mines. Normally you can figure that the amount the world adds to its reserves in any one year is a rather small percentage of the accumulation of gold throughout the ages—relatively little is lost. There have been a few instances in history when this was not true, when the addition was a substantial percent of the accumulation of the ages. This was one of those instances. The Western world's monetary gold increased by almost half in seven years time and probably most of it came out of the ground. Sales by Russia and unrecorded hoards may have contributed a small amount.

FRIEDMAN: No. There is a pertinent distinction between the official Central Bank money holdings and total money stock. Now a large part of what we got did not come out of Central Bank money holdings, but it came out of private holdings as opposed to being newly mined out of the ground, though none of this has any great importance for our present problem.

SCHWARTZ: I have what is the most innocent of questions. We have heard described a series of governmental blunders. Usually, when there is a governmental blunder from one group's point of view, there are gains to another group. Am I correct in interpreting this evidence as a series of blunders that did not benefit any identifiable group in the population?

BUCHANAN: You had a series of blunders, of course, but out of these blunders did come a very substantial increase in the quantity of money. Out of this increase in the quantity of money did come a substantial increase in real income and a reduction in unemployment; so you did have wide-spread gain. Now as to the particular ways in which it was done, people holding silver gained, as well as not only those holding gold, but, as Professor Kemmerer emphasized, those holding gold-mining shares. People in the export industries gained, and prices in general were driven higher.

A large part of this, as Les Chandler said quite properly, was an attempt to drive down the price of the dollar in terms of foreign currency in order to stimulate American exports, and we did. The counterpart to the large gold import was the export of goods and services. That is how the goods were being paid for. If you want a modern counterpart, it is the accumulation of surplus dollars by foreign governments in the last ten years. That is economically a phenomenon identical to the accumulation of gold in the United States in the 1930s. And the domestic interests benefited from it, or thought they benefited from it, as well as the producers of export goods and the producers of substitutes for imports.

SCHWARTZ: My question is really more towards the earlier blunders. Is there no plausible hypothesis that can link the occurrence of the blunders with the group who caused the blunders?

ALCHIAN: What about looking at the political interests? I am admitting to bankruptcy in my political theory, but I object to economists talking as if there were no applicable political theory or that we should not search for one. Now, I do not know how to explain how the stockholders or how the politicians gained. When a business does something and survives, I try to ask what it was that was in its survival interest, whether the acting individuals understood or were conscious of these things or not. And here is a political action, so I think we should approach it first by asking why these actions were taken. Was there a gain to the politician that we have not yet perceived? Instead, we are talking here as if there were some specifiable normative behavior. We seem to be saying that politicians should act in the public interest. That is odd for economists.

I blundered. I was really thinking of the subsequent case. But nevertheless, let's think of what happened subsequently, which you could also treat as a blunder. My point remains that the blunder only means that we do not understand political theory. Now there may be true blunders, but I hate to go around saying, well, it is a blunder, and I just don't understand, and, therefore, there was no motivation, no one's self-interest in the action at all.

DAM: Well, with respect to the '31–'33 deflation, without knowing anything about it, why can't one assume that it was brought on by the Fed, and that the Fed was composed of politicians and very special clientele groups, banks and other creditors, and that the reduction of price level is good for creditors, at least so long as the debts are paid in nominal dollars.

TIDEMAN: But I believe that there just are not that many people who are exclusively creditors. People are both creditors and debtors, and while there may have been some people who would benefit from the overall deflation, most people, even though they may have been creditors, were made worse off by the reduction in overall economic opportunities than they were made better off by the possibility of a higher real value of their creditors' claim.

Now, I wanted to also talk on the point of the difference between the blunder between 1931 and 1933, about which there seems to be general agreement, and emphasize the measures that were begun in 1933. I think there will also be general agreement that there were specific groups that benefited from the way in which the recovery proceeded. In particular, debtor interests and export interests benefited from the abrogation of gold clauses and the devaluation.

MELTZER: I would like to make two points. One is I do not think we want to talk about one single blunder. There were a whole series of blunders. There were certain kinds of mistakes that were made that could be explained on self-interest bases, such as the Smoot-Hawley Tariff, which made the situation worse. There were other kinds of blunders made that were explainable on a self-interest basis, such as the unwillingness—and that is well-documented—of central banks all over the world to come to the aid of the Hungarian banks when they were in trouble, the Austrian banks when they were in trouble, and the German banks when they were in trouble. At each time the Fed would have a meeting, and it would agree, for example in the case of Germany, that while American banks had loaned to Germany, it was not in the interest of the Federal Reserve to do anything about that. That was the responsibility of individual bankers or of the Bank of England.

Then there was a promise implicit in the record, and sometimes explicit, that when it came closer to being a matter of self-interest, as in the case of England, we [the Fed] would do something about it. But, then, in fact it did not do anything about it in the case of England either because it decided that such a course of non-action was in the self-interest of the bankers or in the self-interest of the Federal Reserve.

The hot question about the Federal Reserve at that time is why, in spite of an enormous amount of literature and good banking material which they knew about and referred to, literature which pointed out that the proper thing to do was suspend convertibility, the Fed did not do it. Karl Brunner and I would explain that episode by saying that the Fed really behaved almost exclusively in terms of the real bills doctrine, and that its governors were loyal adherents of that view; it was not a matter of self-interest at all. It was just that they had a theory of their field that was wrong. And in that sense it was a blunder, though not one explainable in terms of self-interest.

NUTTER: I just want to make one simple point along the lines that I think Alchian [pp. 163–64] was moving. The election of 1932 occurred in a time of domestic crisis and the President-elect understood that in order to gain command of the situation and to stay in power and even to increase his power, he had to do something very dramatic. And so there ensued his 100 days of dramatic actions. These various steps that he took, including the calling in of gold and the closing of banks and so on, were all a part of this game plan. He at least had to give the impression of action, of doing something about the situation. And I suggest that the election of 1936 demonstrated that he achieved what he set out to do.

MANNE: What interest did anyone have between 1933 and December 31, 1974, in maintaining the illegality of gold ownership?

TULLOCK: I think there are those who take seriously the prospect of competition for fiat money from some form of private currency. Now in Friedman's paper he said it could not happen unless the government clearly called for a policy of inflation. Well, my prediction is that it will. In fact, I will even go so far as to say it has done it up to this point in time. We see that businessmen always want to reduce the amount of competition they have, even when the amount of competition they have is trivial, so why shouldn't the government? I think this may be noted here.

FRIEDMAN: I do not believe what Gordon Tullock has said

is the explanation. I do not believe they [the managers of the Fed] were ever worried. Maybe they should have been, but I do not believe they were ever worried about the possibility of a private competing money. I think that until after World War II there was simply no pressure on them for doing it. Now after World War II, I believe that they had a vision of a need for a world system; they wanted a managed world monetary system. And they, along with their fellow managers in other countries, believed that it would be easier to manage that system, on the Bretton Woods basis, if they did not let gold get into private hands. They did not want an honest-to-God gold standard. They wanted the Bretton Woods kind of a managed system with fixed exchange rates.

MANNE: But how does that differ from Gordon's point?

FRIEDMAN: It differs, but it is not inconsistent with it. Because it says that even though they thought that the chance was one in a trillion of a private gold currency system, yet they thought that the sensible thing to do was to concentrate the holdings in gold and give the greatest amount of control over that gold to the monetary authorities. That would enable them to manage this international monetary system with the least amount of private interference.

In the same way, I think that the objection of the Treasury to the private ownership of gold is very simple. It could not see anything to gain by it. There were very few people who were anxious to buy gold. On the other hand, they figured this is a bargaining counter we can always use in our international negotiations with other countries to whom we want to talk about gold. But, it is perfectly clear that once we permit private ownership of gold, that will be another nail in the coffin of an official monetary gold standard. It may increase the chance of a private gold standard, but it reduces the chance of having an *official* gold standard. Gold becomes a commodity. You could have it pulled out from under you, and private people can own it, and therefore central bankers switch to SDR's as something else they can really control.

I think we ought to point out that what Karl Brunner was just saying applies perfectly today. In a recent speech by a Mr. [Richard] Debs, the Vice Chairman of the Federal Reserve Bank of New York, he talked about the problem of inflation and he has one sentence, the greatest understatement of the year, in which he says: "The Federal Reserve system is not talking." That is the sum and substance of all the discussions of the role of the Federal Reserve System. It is producing the inflation, but once again economic laws are not working the way they used to, which is Karl's point. And what incentive does the Federal Reserve have, what reward is there to the members of the Federal Reserve to correct their inflationary policy? None.

MELTZER: If you want to put it into institutional terms tied in with both of these topics, the fact is that we have institutionally structured the control of money and intimately linked that with the control of credit. So there is a problem of insulating the two parts politically. We would have had a much better system if somehow you could have separated out the body in charge of the quantity of money and the body in charge of credit. It is because these two have been linked that you have what has just been noted, that you have a constituency that really has an impact on the Federal Reserve. It could be the commercial banks, the savings and loan institutions, the housing market, etc., because the Federal Reserve is operated through the political market rather than in terms of the money system.

III. The Relationship of Economics to Law:
Legal Position of Gold After Dec. 31, 1974

MANNE: What can we expect in 1975 and after?

JONES: I really do not have any strong predictions, but I do not think very much will happen after January 1st because there seem to be few potential American demanders for gold. People have known this for some time and probably it has already been reflected and discounted in the price of gold.

To me the interesting question is what kind of world to expect in the next two years and what will be the role of the United States as a potential supplier of gold on the market? For there is where we stand to make an impact if we try to do anything with our official holdings of gold.

MEISELMAN: I share Professor Jones' view, but there is one other issue that is rather intriguing. That is the question as to whether there will be privately-created gold certificates, which will then become competitive with publicly created money or money that is under public control. This issue was raised by the editor of the *Commercial and Financial Chronicle* several weeks ago. According to him, there are plans afoot to have banks issue essentially warehouse receipts against individual, numbered pieces of gold that they will keep in their vaults, so they will be a receipt for an equivalent piece of gold. It was suggested that these circulating warehouse receipts might become highly competitive with Federal Reserve notes in other forms of what we now think of as money, thus raising the possibility that this private money might drive out public money. I am very dubious that that would happen. It will be an interesting and intriguing experiment, but I really do not think there is much substance to it.

JOHNSON: I have a few points I would like to make. A technical one is that if there is to be a big American holding of gold privately, there would have to develop a market for it. The British market now involves a small group of representatives of companies fixing a price twice a day. Also, there is the question of how they operate a price-fixing arrangement at that time. They get all the orders in by themselves and fix a price that moves twice a day. How would that be coordinated with the market here, which would have to take account of the five-hour time difference with the European one?

The other point is related to something David Meiselman said. I do not really see gold becoming a substitute for ordinary money, given that the price of gold would be somewhere around $175 an ounce. That means that that gold is not going to be a circulating asset, but it is going to be a "hoarding" asset. And that raises the real question of how long it will be before the Internal Revenue Service steps in to make sure that those assets are not used for purposes of tax evasion. The major reason the Treasury has such an interest now is that Swiss accounts and gold are a means by which people can avoid taxes. When you've got certificates, you've got a record that can be traced fairly easily. But gold bullion is simply another matter.

KEELER: I have been following gold for a while and I am doubtful that there will be a great American demand for it. Right now anybody can go buy as much gold as he wants paying a rather small percentage over the face value of it. You can buy one-ounce coins recently struck bearing a 1913 date, produced in Austria and various other places. So the small saver can get gold now. I think the previous comments by Friedman and others were correct. These notes are not going to circulate. Gold can now be used as a savings media by the people who do not trust banks anymore, and there have been gold notes for a long time. The Bank of Nova Scotia issued them some years ago. They were easier than stocks to transfer, as they were not registered. However, I understand in the late '60s, as a gesture of friendship across the border, they ceased issuing them in this form. However, I do not know why others could not be issued in the future.

Finally, I should like to ask why, if there are other countries where you can own gold, it has not become an auxiliary currency there? Why, for example, don't the Swiss use a lot of gold-clause contracts?

MANNE: Does anyone know anything further about this question Davis Keeler raises of why countries where gold ownership has remained legal have not developed the use of gold clauses and gold-backed private currency?

HOLTZER: Maybe because you can get gold bullion.

MANNE: But wouldn't it be a lot more convenient to have certificates, if you trusted the banks, than to be carrying around all those ounces of gold?

DAM: Well, generally, in the U.S., we haven't had any significant use of multiple currency clauses, which I think have many of the same advantages for the people who use them and are, I believe, legal. They are very common in some European countries, especially during the period in which the price of gold has been fluctuating very widely.

HOLZER: I believe that there is sufficient statutory authority to enable the Treasury Department to promulgate regulations and require things like registration of purchases of gold certificates and bullion subsequent to December 31, 1974. Parenthetically, this suggests that this is a good time for people who intend to buy bullion after December 31 to pay a small premium and to buy bullion coins that can move freely without much danger of registration, should it appear. But, more to the point. I have a great concern that subsequent to December 31, 1974, when, I do not know—we could be confronted once again with "illegalization." I say that for a variety of reasons. I am putting aside moral, ethical questions like, "Should it be illegalized?" or political questions like, "Will there be a sufficiently vociferous pressure group to get it illegalized?", and so forth. But if we concern ourselves solely with the legal question of whether it can be illegalized, I think the answer must be an emphatic "yes" for the following reasons. First, I think it is fair to say that historically the original illegalization has stood with no significant legal disapproval; next, in the early '60s, when there were hints, balloons, or bubbles, to the effect that gold was going to be legalized, the Treasury was very quick to deny it, and this quick action was roundly approved by the American Bankers Association. All of the proponents of legalization were virtually stifled.

More recently, you will recall that [Treasury] Secretary [William] Simon said leave it to us and we will take care of it; we don't need a law. Remember, also, that contrary to what was suggested before, the Congress had not clearly embraced legalization willingly. I am sure you are aware that legalization came only because it was finally annexed as a rider on the AID bill that the government wanted passed very much. One wonders what would have happened had it been a separate legalization statute that was not tacked onto anything else.

You may have seen in the press that as legalization came nearer and nearer and finally took place by Ford's signature, various officials of the government, while not explicitly saying they would dump gold, were quick to remind everybody how much gold the government had and what a depressing

effect it might have on the market if they released—I think
was the euphemism—certain amounts of their gold stocks
into the market. Plus, I am not at all convinced that the
legalization statute entirely wipes out all vestiges of prohibition that were embodied in a series of proclamations and
executive orders and in the Gold Reserve Act and in a whole
lot of other things.

And even if it did succeed in reversing all this, I am not
at all sure that existing standby emergency legislation on
the books now could not easily be used as a justification
for again illegalizing it. For example, the Selective Service
Act still contains paragraphs that legally allow the government, after placing an order with an American company
and not getting delivery on time, to take over the factory
or the business. So, I think a host of other emergency
legislation might well serve as a peg for the executive branch
of the government again to illegalize gold and gold clauses.
A couple of other points quickly—one is that the 1950 double
taxation treaty with Switzerland, which was supposed to
have the very benign purpose of avoiding double taxation,
served as the grounds for a case called *X vs. the Federal Tax
Administrator in Switzerland* that some of you may be aware
of. As a result, American tax authorities claiming tax frauds
under American tax laws were able, with the acquiescence
of the Swiss, to get into Swiss bank accounts held by American
citizens.

In addition, there is a treaty that has been signed with
Switzerland, but not yet ratified by the Senate, which is
supposed to deal with the Mafia and organized crime but
which goes a long way toward removing a good part of
the secrecy that is supposed to exist on the part of Swiss
banks. And finally, our own Bank Secrecy Act is a very
dangerous piece of legislation, in that it requires definitive
reporting on the transfer of funds in and out of this country
as well as in and out of your own bank domestically. It

also covers hand-carrying of what they call monetary instruments, including gold coin, out of the United States.

So what I am suggesting is that if for "economic" reasons,
or if for political reasons, an executive department or federal
administration wishes to illegalize gold again in the near
or distant future, they will not find it as difficult to do as
some people naively believe. The machinery is in place and
I think the history of its use and legality is abundantly clear.
So, in answer to the question of what is the future for gold
ownership from a legal point of view, I should say "shaky,
if the government wants to illegalize it again," and I do
not think they would need another act of Congress to do
it.

EPSTEIN: It seems to me that if they illegalize it again, the
government will probably have to pay market rates when
it comes to confiscation. After all, in 1933 the new price
of $35 was higher than the previous market value. Now
it works quite the other way, given a market value of $180.
If illegalization means the government is going to call all
gold in, I take it that the Fifth Amendment will require
that the government should pay that amount. And that seems
to me to be a very strong constraint on such a call occurring.

HOLZER: I think it is very, very significant that when they
illegalized gold in the 30's they made an exception to coins
of recognized numismatic value. And, I think the reason
they did that then is that, had they taken that which could
not be considered part of the money supply or financial
system or whatever, they would have had an expropriation
problem under the Fifth Amendment. But what bothers me
about what price they have to pay is the fact that, to the
extent the government considers, and the Court were to
agree, the gold held in bullion coins is part of the monetary
system, I think they could legally use the so-called official

price now and pull it in at that rate.

LEVINE: It seems to me that, with all due respect, that your reading on the Fifth Amendment and expropriation is not correct. And in the same way I think Professor Friedman had some difficulty with it this morning. I don't know historically why numismatic coins were exempted, but if there was a Fifth Amendment problem it was probably because they didn't think it was worth paying the numismatic value to get the quantity of gold that was represented in that form when you could get the major proportion of the gold stock in simply by paying the existing official market price. Under the circumstances, when the price was about to be raised by the government, that expectation value is not recoverable under the Fifth Amendment taking clause. Reversing that, a situation where there was an existing market price for gold well above a specified monetary price, I think the government would have great difficulty under the Fifth Amendment arguing that it could pay less than a legal market price. Where there is a legal market price, that would certainly be the point of departure for any compensation that had to be paid under the Fifth Amendment, I cannot imagine that the government could find a way to convince the Supreme Court, not that a new illegalization of gold was not all right—that's not a problem—but that the government did not have to pay compensation at something resembling a recognized market value at the time of the expropriation.

HOLTZER: I'm not disagreeing and I didn't say they'd have no problems. I merely suggested that the government could argue that "we told you the official price is forty," no matter what it sells for, or what they want to do with the London prices they were fixing, or what people want to pay to banks for bullion coin. What price Chase wants for wafers across their window has nothing to do with us; we said the official price was forty some dollars, and we're calling it back at that price. Now how do you answer that?

MELTZER: Was anybody ever prosecuted for failing to turn in gold?

HOLTZER: You're damn right; there were confiscations and civil penalties and forfeitures and use of the whole arsenal of penalties that exist.

MANNE: The crucial point here, addressed in the Buchanan-Tideman paper and in Friedman's paper, is the question of the extent to which the Supreme Court would be willing still to attribute monetary characteristics to gold, because only then could they bring it under the money value clause of the Constitution, thus adding strength to the Holtzer position and probably weakening Professor Levine's constitutional view. Does anyone think the government would have to remonetize, i.e., coin gold to bring the clause into operation, and, if not, with what effect on any price that would have to be paid upon a new illegalization?

SANDALOW: Every once in a while a local government will adopt an ordinance zoning land for park purposes only, or for municipal building purposes, and then try to condemn the property at the now lower value. They don't get away with that, and generally it is thought to be a violation of the Fifth Amendment.

BRUNNER: It seems to me that it would have been better than what the Federal Reserve did than to do nothing whatsoever.

JOHNSON: I'd like to say in the first place that the arrangement for suspending convertibility of bank deposit notes is not the same thing as changing the value of deposits.

One other point that I want to make is that we should start looking at some reasons why governments confiscate gold, independently of the manner in which they do it. The major possibility I can see now is not the fluctuation of a fixed exchange rate, the possibility which may or may not be considered so unlikely with the Arab countries taking all this money out of western nations by the sale of oil. Rather, there may be different uses of gold. It might be the case, as happened in World War II, when we shipped a lot of gold to China to help them rescue their government by stabilizing their currency. We gave them gold they wanted to back their money. Now the U.S. might itself want to raise a lot of gold for some purpose connected with war or the prospect of war in the Middle East. If other people value gold and don't value paper money, you may want to have it for military purposes, in which case you might want to take it away from our own citizens to mobilize your strategic defenses.

MANNE: Is there anything to the argument, or to answer Warren Schwartz, that to the extent that governments fear that gold serves as a scoreboard for inflation, they just don't like having it around?

JOHNSON: Yes, but I don't see that gold itself now is that important. You can keep talking it up just so much.

MANNE: But that leaves the question of whether individuals will flee to gold if they fear inflation domestically, and whether they do so rightly or wrongly, it may have the political effect suggested.

JOHNSON: I just can't see it as a popular hedge for Americans, as I think we'll see after December 31.

LEVINE: While I agree with Winter and Epstein that a mere sham remonetization would not justify the calling in of gold and the paying of some artificially low price, there still may be some reason to do it. There are, as Mashaw says, takings and there are takings. If the government gradually imposed a set of controls, price controls, for example, on gold, and ran that for a while in such a way that it did not become a taking, you could gradually black-marketize all transactions at above the official rate. You would have to do it for a bit of time, but you could imagine a scenario in which the government remonetized gold as part of a full control program, allowed the price to rise in a black market, and then called in gold at the controlled price. And, if they took long enough to do it, and if they made enough noises about social reasons for doing so, that would very likely not be a taking.

IV. The Relationship of Gold Clauses to Indexing Provisions

WEISTART: I would like a clarification as to why you would use gold as your index device now rather than something else. It seems to me even if we say the gold-clause case still stands that there must be other contractual devices for accomplishing the same purpose.

KEMMERER: I think I made a comment on that yesterday to the effect that a great many people have a tremendous faith in gold and they see it as something that they understand and in their way believe in. The point, I believe, is basically that it is a matter of personal preference. It isn't necessarily the best index, but it is the one most highly regarded.

MILLER: If you want to make a prediction on what businesses would do to reduce the risk of changing price levels, you'd ask what commodity or what type of index would they use, and I don't think that you would predict that very many people would use gold clauses. The price of gold certainly has been less stable than the CPI in terms of predictability. If you wanted something that fluctuated like pork belly futures in your contract, you might want to pick gold, but otherwise, as an index, there are better ones than gold.

KEELER: Speaking of multiple currencies contracts, I think countries other than the United States have the option of using the more or less stable dollar if they don't like their local currency. On the other hand, in the United States we don't have that option.

MELTZER: That sounds wrong to me. Individual currencies will fluctuate and presumably the most stable unit will be a basket of currency. As a matter of fact, that contract already seems to be developing. It is hard for me to understand why anyone would say, we'll take our risks exclusively in the form of dollars when the risks can be reduced through diversification, just as mutual funds do with stocks. I believe Barclay's Bank and several others already offer a bond contract on the European bond market, denominated or payable in a basket of currency. The basket of currency will have changes in prices, but it will be more stable than any single currency, chosen in anticipation of shifts in relative prices.

TIDEMAN: It seems to me that the only time somebody would want to enter into a contract where payments depended on the future price of gold was if he had a series of business deals where that constituted a hedging agreement for him because he had to purchase gold for non-monetary uses, such as a dentist would.

EPSTEIN: It seems that the answer to that is that he would

simply go into the gold futures market and not be tied to any other contract provision. That is, people worried about the future price of gold would buy gold futures.

JOHNSON: I'd like to try to explain the "basket of currencies" notion. You've got two different things to be concerned about: stability in terms of money and stability in terms of the purchasing power of goods. This particular deal is designed to give you stability of purchasing power in terms of monies by getting a weighted index. You're investing in assets in your non-money market, and you have a weighted bundle of currencies in terms of which to express the contract. But it doesn't tell you anything at all about the purchasing power of commodities. But the purchasing power of gold over commodities is very likely to be far more unstable than any particular currency you can think of.

HOLTZER: I just wonder how the economists would compare gold, which sold for so long at $35 an ounce and now is at $180, with the so-called basket of currencies. It appears to me that the gold was the better hedge.

MELTZER: If you compare sugar over the last year with gold, then you find that sugar was a better protection against inflation than even gold. All you're saying is that if you look back now you find that you missed a great opportunity. But that gives you no information about what is going to happen to the price of gold versus commodities, or the price of gold versus currencies from here over the next similar period, or whether you would rather be in gold or sugar today.

HOLTZER: Of course, that is what a lot of economists said when gold was $35.

MELTZER: That's true. And that's what a lot of people said when sugar was ten cents a pound.

ROGGE: I really am somewhat disturbed by the casual way it seems to me that we have dismissed this question of the abrogation of the gold clause in contracts both public and private, and seem to be going on to other issues. As a matter of fact, it seems to me that we have in a sense been joining our voices with the unwashed masses who always cheer when Shylock is denied his pound of flesh. And we may cheer lustily when we see his pound of flesh being eroded daily by inflation; but still my own feeling is that the decision was morally and economically wrong. And, it was constitutionally wrong in the very special sense that, had I been on the Supreme Court, I would have found it unconstitutional.

But, I'm disturbed by the justifications that seem to have been accepted here. Windfall gains and losses are the stuff of life, and in particular the stuff of economic life. To say that these were very special windfall gains and losses brought on by government action and that, therefore, to have insisted on the performance on the contract would have frustrated the intent of the contractors seems to me to have been wrong in many cases. It is a matter of fact that many people used gold clauses in their contracts for the explicit purpose of protecting themselves from government, from the capricious and arbitrary action of government. As a matter of fact, what did happen was precisely what they were trying to protect themselves against.

Miss Schultz has here a contract that Mr. Pierre Goodrich, who founded the Liberty Fund, drew up in 1927 that very clearly had as its explicit purpose to protect himself from capricious action by government. So for us, then, to argue that because this was a change occasioned by government

action the abrogation was justified seems to me to be denying
the very basis of contract. And, as a matter of fact, if we
accept that, then we have to accept anything the government
does from here on out; and I see no way in which any
of the schemes any of you are talking about here can be
any more successful as a barrier against government confiscation of property through inflation and in other ways than
it has been in the past.

Our decision as to what is the best way to hedge against
inflation, it seems to me, is not one to be made in this room
anyway, or even in the chambers of the Supreme Court,
but by the freely-acting individual who can choose, if he
wishes, the alleged inferior contract, the gold contract. In
other words, it seems to me that again this is a choice that
should be left to the individual, and I have not as yet heard
any justification for society intervening in that choice as it
did in the 1930s.

DIETZE: I should like to elaborate a little bit on what Mr.
Rogge just said. We discussed yesterday the constitutionality
of the abrogation of the contract clauses. And, I regret that
the word constitutionality was left a little vague. I think a
very clear distinction should be drawn here between constitutionalism in the philosophical sense in the sense of modern
liberalism, and constitutional law in the sense of the American
Constitution. There is a big difference between these two.
Constitutional government in the philosophical sense is limited government, but that does not imply at all the absence
of government and absolute freedom. The limited government idea is very characteristic, for instance, I think, of
the first great author on constitutional government, Montes-
quieu, in his main work, *The Spirit of the Laws.* He fought
despotism. He fought a maximum of government interference, and yet he made it quite plain that government should
exist and be limited. Both Tocqueville and Adam Smith made

it quite plain that there could be really no free market without some kind of government, without some kind of government regulation. In other words, liberalism and constitutionalism are absolutely compatible with law.

As a matter of fact, I should say that law and a certain degree of government control are prerequisites for a free working, liberal society, and it was so conceived by the great liberals. And I would add, in complementing what Mr. Rogge said, that one of the important principles of classic liberalism has been the idea of free contracting with the contracts binding. One of the main functions of the government in a liberal state is really to see to it that contracts that are concluded freely among individuals are binding and are being observed. And, I would say even from the point of view of liberalism in general and constitutionalism in general that the abrogation of the gold clauses was a very dangerous thing because it destroyed one of the most important principles of the law and of liberalism, the old principle, *pacta sunt servanda* [contracts are to be upheld]. This is my answer as far as constitutionalism in general is concerned.

Now let me turn to constitutional law in the sense of the United States Constitution. Edward Corwin, probably one of the best observers and commentators on the American Constitution, in one short piece about the progress of constitutional theory from 1776 to 1789, emphasized the importance of the recognition of the protection of contracts by the government. He actually said that the main reason the American Constitution was adopted was because of the interference with contracts on the state levels by majoritarian governments, and he said one of the most important clauses of the American Constitution is the so-called Contract Clause, the clause saying that no state shall pass any law interfering with the obligation of contracts. This is the very gist of the liberal American Constitution and, as you know, the Contract Clause was interpreted in a very generous way, by John

Marshall. When the Contract Clause no longer was considered sufficient to protect property and contracts, these institutions were protected by the Due Process clauses. But after 1933, when the gold clauses were abrogated, this was a serious interference with contract.

Now you will answer that the Constitution only stated that no state should pass a law abrogating the obligation of contracts, but I would say that even more so, perhaps as a matter of due process, the national government should be prohibited from passing laws interfering with the obligation of contracts, which, I think, regrettably was done here. And I think that decision really pulled the carpet from under the American Constitution. I think this is really where our present troubles started. Once you do that, once you tell people they need no longer fulfil their contract, you run into enormous problems, and I think the origins of the present law and order problem date back actually to this first abrogation. The Supreme Court caved in before the antiliberalism of the New Deal, not only in 1937, but already in 1934 and '35 when it decided that the laws providing for the abrogation of the gold clauses were constitutional.

WINTER: I don't know whether Ben Rogge was trying to prove that Warren was right, but it seems to me that you've missed not a legal point but an economic point. I think you are confusing hedging contracts in a market trying to give some kind of predictability, some kind of protection in the price of a particular commodity, and a clause in an independent commercial contract trying to give some kind of predictability as to the value of that contract.

If you interfere in the hedging market, I would suppose then you are doing what you described. If you interfere in the other market or you interfere in individual commercial contracts, you have that effect only if the interference is such as to bring upon one of the parties the harm against which they were trying to protect themselves. The harm here against which they were trying to protect themselves was a depreciated currency. That's what they intended. At the time the clauses were abrogated, the currency was in fact appreciated.

Now, you may perhaps show, and it may well be, that this should have been left to a case-by-case adjudication using legal doctrines like frustration of contract or something like that, but it doesn't seem to me that in the particular case you are talking about the intent of the parties—and that's what freedom of contract is all about; it's not just the language; it's the intent of the parties—really was abrogated.

Now, as for Mr. Dietze, the history, so far as most people have been able to determine, is that there was not all that much focusing on liberty of contract or the like by the founding fathers. Their particular problem was trying to establish a nation and trying to establish a common market. Maybe you can reason from that to the other principles, but it does seem to me to be a bit of a jump.

Also there was not the history underlying the due process laws that one expects. The best view is that the phrase refers to a government of law as established in something, e.g., the Magna Charta. If you go back there, they define the government of laws as one with due process.

Finally, it seems to me the whole upset about abrogating the gold clause is based on language, as I said, rather than intent. So when we come to the contract clause, which reads that no state shall impair the obligation of a contract, the language isn't really important anymore.

BUCHANAN: Well, I'll associate myself with much of what Ralph Winter says here in response to Ben Rogge. The point was, it's true that the gold clauses were designed to protect against the capricious action of government, but it was the capricious action of government in inflation that they were worried about, not the opposite. And this created a wholly different situation. Now, I think you can make an argument that the government should not have devalued; they should not have called in gold. But that is not what positive constitutional law is about.

To go back to Gottfried Dietze's point, I agree that it is a tragedy in the sense that what happened is necessarily our constitutional law. But I don't see how the Court could have done anything else, given the situation it found itself in with this tremendous burden that would have been imposed on the debtor interests because of the deflationary situation.

MANNE: I'm going to exercise a prerogative of the Chair for the moment to second much of what Rogge said because I think that both Buchanan's statement and Winter's overlook the extent to which the government itself was responsible for aggravating a situation to the point that the Court felt that it had to uphold the constitutionality of this because otherwise there would be great windfall gains.

It seems to me that a holding in the opposite direction would have been the kind of discipline on the powers of government that were exactly what the Court should have been interested in establishing at this crucial point in history. Just like freedom of speech, it becomes really significant when someone has said something very bad that you don't want to hear.

KITCH: If you're worried, as I think everyone should be, about such things as modern price control or oil price regulation, developments which involve massive abrogations of contractual undertakings, you don't need to look to the gold-clause abrogation cases to find the constitutional precedents that are troublesome or defective or wrong. There are other principles at work that lead to far more massive disruptions of contractual relations than these particular cases. And I certainly share the view of the principal paper that these cases are easily explainable on rather elementary notions of traditional contract law.

I do think, however, that there is something very unsettling about the events leading to the gold clause abrogation that has not been focused on here and that I think should not be ignored because it's had a very profound effect on actual policies of the national government. These are the aspects of that series of events involving executive constitutional style, if you will. That is, Roosevelt's "Hundred Days"—and these actions were among the first—have been very important in modeling the notion, both in the legal and political sense, of what Presidents ought to do.

First, you have here reliance by the executive on ambiguous clauses in long past enacted emergency powers legislation. This has become a fount of great executive power in the United States.

Second, you have the general notion that has developed out of these events of the importance of dramatic and precipitate action by the executive taken prior to Congressional deliberation on the grounds of urgency, when often the action taken cannot be reversed. Of course, Roosevelt went through the process of subsequent Congressional ratification of these steps, but the legality of his actions was seriously challenged there. In subsequent developments in the executive branch, we've even seen the dropping of any effort at obtaining Congressional ratification. One notable recent event is the reliance on emergency powers for the imposition of the ten-percent import surcharge as part of the great "announcements" of August, 1971. The Court of Customs and Patent Appeals has recently held that surcharge illegal and in excess of the powers of the President, a holding which reflects the beginning of a judicial trend away from these notions of executive power that have been so important for the last forty years.

But going right back to Roosevelt's "Hundred Days," the executive has been expected to exercise and has claimed for itself the power to engage in far-reaching measures without explicit Congressional authorization. The fact that Roosevelt successfully completed this series of maneuvers is very important as a precedent for that. And I think that it has created very serious problems for the principle of separation of powers and for the authority of Congress. Even to this day, it's a little hard to figure out exactly what was being accomplished by some of these "dramatic actions," but there was an idea that it really didn't matter as long as action was taken and it was dramatic. Then it was socially desirable and useful and politically good, and this preference for action without regard to its consequences—which has certainly become a very important part of the general popular political culture—has plagued us to this day, and these are things that I hope we are now beginning to reconsider. So this series of actions relating to gold is central to that whole set of constitutional issues, and I don't think we should overlook how profound and serious they are.

V. The Monetary Constitution

KEMMERER: Going back a bit, I have three quick points I would like to make in response to comments made by Ralph Winter and others. One is that the purpose in 1934 of the devaluation was to raise prices. They hoped to raise it back to exactly where it had been in 1926, but there was no absolute certainty that this would be accomplished. The purpose in the 1890s at the time of the Bryan presidential campaign was to right the damage that had been done during the preceding generation. At this time there was a great increase in the writing of gold clauses. So, in '34 you had the situation for which they were being written, for which they had been created. No wonder that many people felt that here was a situation they had looked forward to and should benefit from. That is point number one.

Point number two is that in the 1780s there was a great deal of worry about abrogation of contracts—Shays' Rebellion involved that. Probably Shays' Rebellion is considered one of the major reasons for the Philadelphia meeting that resulted in the Constitution itself. So don't overlook this background in terms of worry over abrogation of contracts.

And point number three: At least down to the Civil War, the statement, "No state shall make anything but gold or silver legal tender," at least in practice, was interpreted to include the federal state as well as the other "states." Now after the Civil War, the various legal tender acts changed that situation, but down to that time the federal government did not create any legal tender money, and it felt the same obligation, apparently, that the Constitution had imposed on the others.

MELTZER: One brief question I would like to ask the lawyers. What would have been the situation if the contracts had explicitly said that the clause was not designed as a hedging device?

MANNE: Professor Epstein.

EPSTEIN: I think that the clear answer, if they said, "Look, we want this clause to mean what it states, and we're not dealing only with a hedge because of deflation of currency, inflation of currency, depreciation or appreciation," then at that point you would not be able to use the argument based on contract construction.

MANNE: But the Supreme Court was not interpreting particular contracts. It had a provision of Congress before it. And, if any one of them had been shown to have the language you mentioned, would that have made a difference?

EPSTEIN: There are two separate questions: the one that Henry Manne put to me and the one that Allan Meltzer put to me. The question Mr. Meltzer put to me is, suppose you found a situation where the contractual language was explicit and that the contingencies to be guarded against were stated. The question would then be, could you use the doctrine of frustration? And my answer is that the more specific the language, the more difficult the frustration argument would be to apply. Under those circumstances you would then come to the freedom-of-contract issue, which I think in these cases really was not raised.

The point Mr. Manne made is that in this particular case the abrogation was done by statute, by Congressional action, and I fully agree that, indeed, that was the mode which was used to handle the particular problem. But, one could argue—I think quite sensibly—that when you have a million separate transactions, all of which assume a common form, that it is cheaper administratively and it is conducive to the contractual certainty to make explicit by statute what you are going to do. The vagaries of state by state litigation impose too great a wait.

There are lots of times when common-law actions are not successfully enforceable by traditional private-law remedies, occasions on which the state—and I think in a thoroughly proper fashion—decides to standardize the remedies in order to remove the pressure from its own judicial system and the uncertainty from the general administration of justice.

For example, we might have all sorts of private actions to control judicially automobile emissions in Los Angeles. But it is too costly, too uncertain, too silly, and so what the state does is to try to impose some sort of regulation. The state may perhaps restrict the freedom of contract (or at least of action) by saying that you will have to pay a certain tax if you wish to emit a certain pollutant. And, it seems to me in general that those kinds of things are going to be important in these cases.

So, in answer to the question in this case, I think you don't have much of a problem with freedom of contract; in other cases you might. And, indeed, I think Mr. Dietze has made a very fundamental error by trying to make the gold-clause cases very important for the law of contract viewed as an aspect of political theory.

Let me analogize to the law of private corporations, which I think shows something about the very special place of money in the economy and why it is that one ought not to generalize from money cases to anything else. Let us suppose that we have a corporation that has a hundred shares outstanding. I have entered into a contract in which I agree to sell my ten shares of the corporation to Mr. Y for a stated price. Then there is massive redemption—50 percent of the shares are called back by the corporation. Now the buyer comes to me and says, "Epstein, the contract calls for the delivery of ten shares of the Corporation; give over those ten shares." But now I only have five shares. "Those," Y says, "are mine; you give me the five shares now; you pay me damages for the value of the five shares that you're not delivering." Mr. Dietze, I submit to you that there is not one court in fifty states, in England, in the British commonwealth, on the continent, in Soviet Russia, or wherever you wish, that would say that would enforce the contract for ten shares even though it stated ten shares.

What everybody would say is look, in this particular case there was an overriding intent that you were buying a one-tenth interest in the corporation expressed under certain circumstances in the particular language that then conveyed that intent to buy a one-tenth share in that corporation. Now when there is a redemption by the corporation not a party to the contract, the way in which we effectuate the intended terms is to give you my full interest, my ten percent, my five shares, and my five shares only.

Well, now the government happens to be the fellow who's running our corporation and it doesn't print shares. It prints money. And it is a very capricious sort of fellow, and what it does sometimes is print too much (just like creating a stock dividend) or print too little (which in effect creates a redemption, as in the case just given). What happens when you've got a private situation between the two parties is say, "Look, to the extent that we could possibly filter out the inconvenience caused by this corporate head, we will read the contract to refer to fractional interests of the total worth in the society not to a particularly number of shares." By analogy, we will knock out the gold clause. And it seems to me that unless you're prepared to say that the stock redemption situation is one in which you are going to give damages, then you have to say that the abrogation of the gold clause is absolutely correct.

Now, one other point. Mr. Dietze, you dwelt at length on all the stuff about *pacta sunt servanda*—contracts are to be observed—then omitted to translate the rest of the Latin learning, *sic rebus stantibus,* which means except when it is unjust to do so because circumstances have changed. It seems to me that if you are going to talk about the law of contract, you ought to talk about both sides of it. We may put aside fraud, duress, mistake, misrepresentation, that from time to time excuse contract performance and note simply that many times frustration-related doctrines in themselves have been quite sufficient. If we're going to incorporate the enforcement of contract tradition as a Constitutional law, you can't be selective and ignore those hedges against over-enforcement that have traditionally been part of the law in both common law and civil law traditions.

So, you really can find a better example than gold clauses to show a belief in freedom of contract; it's got to be a more balanced concept than has been stated here. Pick a good target: attack the minimum wage. It seems to me to make a much better case.

NUTTER: I want to follow along somewhat the same lines. I want to join Ben Rogge and Gottfried Dietze in lamenting what has happened to Constitutional law, but at the same time take the position that this is not the issue on which to fight it. Now it seems to me that when the Constitution was framed one of the primary concerns was to limit the power of the government to define property rights in one way or another, and there are a number of clauses in the Constitution that do this—for example, the contract clause, due process requirements, and, I would add most fundamentally, the taxation clauses. When I raised the question, what is the difference between taxing away property and expropriating it, the lawyers were not quite so certain in their answers. And I would say that if we were to get to the real core of the problem that we face, it is the fact that there are no Constitutional limits to taxation by government anymore, and hence the government has absolutely, totally unlimited power to take away from my property in any way it wishes and to abrogate any right it wants to on the basis of these various principles. So it seems to me somehow that we have to get down to the basic question: Are there to be limits to the power of government to define property rights?

MANNE: That's a good point, which I think can be elaborated a bit in answer to Dick Epstein. It is very easy to make this argument in terms of individual contracts, but it becomes very different when the argument is wholesaled into a constitutional argument such as Ralph Winter does or as Dick Epstein has offered here today. The problem, as Warren Nutter suggests, is that we are left with no lines to guide us on the constitutional questions. The traditional contract questions are fine because they inevitably come up on an *ad hoc* basis with two parties involved. But that is not the constitutional issue, and when Ralph Winter in his paper gives away the whole issue of economic regulation, as I think he does, and as Epstein does with his "convenience-of-enforcement" argument, it leaves us with the question of whether there's a line anywhere. And I don't believe we have any satisfactory way of defining the limit short of the way we do it with speech and religion in the First Amendment.

LEVINE: I want to endorse Professor Nutter's remarks and also endorse Richard Epstein's demurrer. That is, I would like to admit the truth of Nutter's remarks but to suggest that they don't constitute a contradiction of what Richard was saying. I think that the issue Nutter raised is a terribly important one. It is not quite correct that the power to tax is unlimited, but it is pretty nearly correct that a clever government can tax to the point where it reduces very substantially the value of the object of the tax, in a way that is not recognized by the taking clause of the Fifth or the Fourteenth Amendments.

It is also true that through zoning and other kinds of regulation, the government takes, without describing the act as a taking, property rights for which it doesn't compensate. The relationship between regulation, taxation, and taking is a terribly important issue and one on which we might have an interesting conference sometime. The ingenuity and variety of instances in which the government has managed to take, redistribute, and do whatever it likes to interests which we would ordinarily describe as property rights— without anyone's explicitly recognizing what it was doing as "taking," or without being subjected to the political discipline imposed by compensation and other requirements—is a phenomenon very much worth examining.

The problem is that the *Gold Clause Cases,* in my opinion, don't present that issue. First of all, the question is what property rights were being taken. Property rights tend to be established or recognized by a contract, and then exchanged. One of the purposes of the contractual doctrines that Epstein cited so ably is to help the courts find out what the parties had when they started and what they intended to give to each other. If our reading of these cases is correct, what the court was doing was deciding that no property rights were being taken precisely because the arrangement between the parties did not contemplate this kind of situation. Thus, to the extent that it could determine the purposes of the arrangements between the parties, the court was effectuating them more truly by abrogating the gold clauses than it would have by recognizing them. The court was not taking property rights, it was determining that the parties had not intended that there be any under the circumstances then prevailing.

With respect to the government's calling in of gold, the relevant clause aside from the monetary clause is the taking clause of the Fifth Amendment, and the relevant issue is the amount of compensation it required. But, with respect to abrogation it isn't clear that a major constitutional issue was involved at all. I suggest, in other words, that we should not confuse the terribly important question of what the government can do to us and what the Constitution ought to say about that, with the less important question, what ought contract law to be with respect to certain rights and to certain exchanges. The latter is a matter typically litigated in both state and federal law, and federal law even recognizes the supremacy of state law in those areas. So I think to raise that issue to the level of a social contract issue is to magnify it out of proportion.

HOLZER: If the Supreme Court was concerned to decipher

contractual language and to assay contractual intent in the *Gold-Clause Cases,* what was all that public policy language in there about?

WINTER: I would say that, given the validity of everything that had been done before, of everything the government had done up to the abrogation—and I think that was what was most bothersome—the abrogation was reasonable. It was just a reasonable extension of all that, because now those clauses can't be enforced according to their own intent. Hank [Holzer], you know that lawyers use the words, "public policy," to mean anything.

MEISELMAN: I'd like to raise another consideration in terms of the kinds of contract that we're talking about. Many of these, as I understand it, were long-term bonds that were highly negotiable and that were traded frequently, so that a bond, one bond, may have been issued in the year 1900 and may have been traded 30 or 40 or 50 times up to the moment when the gold clause was abrogated. Now it seems to me that most of the discussion that I heard seems to assume that you have some kind of instrument that was not traded, since the question the lawyers have addressed is what did the parties to these original contracts have in mind. I'm not sure that's applicable, especially if you consider an individual who may have purchased a bond or large numbers of bonds in 1933 precisely with a view to collecting on the gold clause.

ROGGE: Dick Epstein has viewed the Supreme Court's decision as one viewing any long-time arrangement as being actually stated in terms of some percentage figure of the national money supply. Now he didn't say whether he meant M-1 [currency plus demand deposits] or M-2, [currency, plus demand deposits plus time deposits in commercial banks], or what was included, and thus he provided us, in his otherwise intriguing argument, with no way to measure the number of dollars that should be paid, though that computation is quite simple in his private corporation analogy—deceptively simple. Let's say that in 1927 Pierre Goodrich drew up a 25-year lease in anticipation of inflation. Had the contract been permitted to run in the '40s and '50s, he would have been protected by that contract against exactly the inflation that he was concerned about and for which that clause was inserted in that contract. Therefore, it seems to me that it *is* a denial of his intent, and that in fact the government deliberately wanted to contravene his intent.

LEVINE: On Dave Meiselman's very interesting point about negotiation, there is a problem of intra-personal equity. The real question is what was purchased when that bond was purchased, and there is really an equity problem between the debtor and the creditor succeeding whoever made the original agreement. What you have is a situation in which the classical legal position on that would be essentially that he did not get what he thought he got and would be damaged thereby. Now if you're going to worry on straight equity grounds, that's between him and the debtor, and you have a real problem there, a very difficult question to answer.

MANNE: That raises another interesting issue, too, and that is whether there was any evidence immediately prior to the rumors that Roosevelt might engage in this behavior, that people had begun to act as though their wealth position was being considerably increased, because that defines the real contract issue.

KITCH: That is interesting. If you're worried about the traded securities, then it would be obvious to look to see what their market prices were prior to the abrogation. You would immediately have a problem because a lot of the debtors were of very uncertain credit worthiness, and I assume a lot of these securities were trading at deep discounts because of the whole probability of receivership and so on. So, what you would want to find out is whether the securities of a clearly credit-worthy obligor were being traded at prices giving a premium to the gold clause. Now the most obvious credit-worthy obligor with gold-clause obligations outstanding was the United States government.

I think we should make it clear that the Court was unequivocal in stating that the abrogation of the promise was an unconstitutional act of the United States government. The Court was unanimous in that. The issue in that case was the issue of damages.

Now one of the ways to have shown damages would have been to show that you had purchased a gold obligation in reliance upon the undertaking in question at premium price over the non-gold dollar price. But I take it if we accept the information that Milton Friedman advanced, that is, that prior to the rumors there was no expectation of a change in the dollar price in gold, then there was no trading premium for the gold clause up until the time the rumors began circulating.

At the time the rumors began to circulate, I would suspect without looking that the market was aware that the package might include some impairment of the right to ride on the revaluation of the dollar price of gold. So I would bet that you in fact did not have a group of purchasers who paid a premium for a gold clause and that's one of the reasons that the Court said there were no damages.

KIRZNER: I want to deal with the question of our monetary constitution. There's one aspect of the Buchanan-Tideman paper that was fundamental, I think, to the thesis of the paper itself, and fundamental to a lot of the discussion that has followed, which I would like to question. I'll read the passage:

> There are genuine public good characteristics of money. Agreement among persons concerning the thing to be used as money will greatly facilitate economic exchange. The delegation of the authority to define the monetary unit to the government seems to be a plausible outcome of a constitutional contract. The definition of what is to be treated as money in the society does not seem much different from the definition of what shall be the ordinary standards of measure and it is no surprise that the power to fix weights and measures is found in the same sentence of the United States Constitution that provides for regulating the value of money.

Now here we have a questionable theory of money. The argument is that in view of the public goods character of money there is an implication that there would be a unanimous social contract assigning to government the power to fix the monetary unit, to define the monetary unit as it defines and fixes weights and measures.

It seems to me that the classic public goods reasoning fails with respect to money and to a lesser extent fails with respect to weights and measures. The first objection—then later on I will come to a second objection challenging the public goods reasoning in general—is that of national defense where, let's say, we all wanted to defend ourselves against the Russians and we all agree that unless we all get taxed there won't be enough money to defend ourselves. Then we can talk about a unanimous social contract that will give the government the right to axe us over the heads and force us to pay the tax as necessary to defend ourselves. The argument there is that if I pay I will achieve what I want to achieve only if you pay, too. Therefore, I agree to be forced to pay in order that you be forced to pay.

That reasoning does not apply in quite the same way in the case of weights and measures, and certainly not with respect to money. And I would argue that there are no grounds here for an implied social contract giving government the right to define the monetary unit. The reasoning, I think, is as follows: It is not that I must use certain weights, or that I must use certain monetary units, but, if I decide to trade or if I decide to measure something, then I must use a certain measure or a certain monetary unit, and I would be able to do so successfully only if others do, too.

Now, there is, however, a way out of this. I don't have to measure at all. I don't have to trade at all, and consequently the public goods reasoning which assumes that the state can force, can compel a particular object or item to be used as a monetary unit fails here completely. After all, money is a question of confidence, and the government cannot say you must have confidence, and that in order to make sure you have confidence we'll make sure that everyone else has confidence, too.

All the government can actually do is say you must *act* as if you have confidence. But I don't have to act as if I have confidence in the money because I can barter, I can circumvent the compulsion to use the monetary unit. There could then be automatic market evolutions of monetary forms that would in fact circumvent this particular type of government compulsion. Now, so much for the argument about the public goods reasoning in this particular case.

Let me turn for a second line of argument to the whole public goods reasoning altogether. The public goods argument takes the assumption that we already know exactly what is required. We already know how to defend ourselves against the Russians, and we already know that if all of us will pay the taxes, we will get exactly what we want. In many cases, however, the knowledge of what is required is itself revealed only in the give and take, the trial and error processes, in the market system itself. So, for us to assume that there is an implied social contract giving to government the power to define the monetary unit implies that we know that the government will know exactly what the desirable characteristics of money are and exactly which particular entity most desirably possesses all of those monetary characteristics.

In fact, the development and the evolution of what money is has itself been a trial and error process that cannot be assumed to be known or have been known in advance. This applies to many cases of public goods, and it seems to me that it applies to money in particular. On these grounds, it seems to me that the reasoning that the government has an implied Constitutional right to define money is in error.

EPSTEIN: I think at this point I would actually like to defend Buchanan and Tideman, at least on several points. It seems to me that the arguments that Mr. Kirzner made really cut in the opposite direction. The important point about national defense is that we could not opt out of the system even if we wanted to, and that once we decide to defend ourselves against external enemies, everybody has to participate or there will be some real problems of internal equity.

The great beauty about money is that no one need be made to use it as the standard of his exchanges.

Nonetheless, it's generally very convenient to have a common standard of weights, measures, and currency. People who don't like it don't have to use it. There is absolutely nothing to be lost by having a common measure because it is so easy for all of us to find a way to do without it in the proper transactions when we so choose. So at that point it seems to me that the reason that we put the power to define money in the implied constitution is because by putting it there we still leave open a large number of private options for transactions not using the common currency. And it seems to me that, given that particular feature, we should like to have the Constitution look the way in which Buchanan and Tideman stated.

The other point I think is this: The theory of implication

used in their paper is a very thin one. Generally, when one talks about contract law, you talk about inference from customs, from the way in which people respond in like situations, and so on. And in this particular case you can look at the number of nations that, in their own constitutions, do not have a centralized authority to administer the common currency of the realm. There are very few which have that particular characteristic, but even under the most stringent interpretations of the minimal state you find a monetary currency available. So, this seems to me a strong argument in their favor to say that if you have almost every on-going legal system, regardless of its general political characteristics, providing a common monetary currency then it is very appropriate to say that the phenomenon comes within our implied constitution as well.

TULLOCK: As a matter of fact, the most successfully existing economy, which is Hong Kong, has no common currency issued by the government. It's not very obvious exactly how the system works. It requires banks to actually put the money up. Basically, however, all I wanted was to object to the public good definition. "Public good" actually is simply a case of a very large externality. We don't observe pure public goods. Where you have a very large externality, the market gets very expensive to use. Under the circumstances, there are efforts to find something else to take its place.

Unfortunately, the government seems to be working poorly, too, and you don't have any clear decisions in any of these areas. What you have are two imperfect instrumentalities. Imperfect market and imperfect government. You have to make your choice between them in any given institutional state. It's not obvious to me that putting this in the hands of the government is correct. Perhaps what we need here is a better designed government agency, which, of course, a lot of people have talked about. The history of those places where money is uncontrolled is pretty bad, too.

KITCH: I want to make some general comments about indexing. I don't think Goodrich in his long-term lease could have told the court that he would have insisted on an index clause if he had known he couldn't have a gold clause. In part, I presume, if he had had an index of the type we generally talk about, it would have depreciated the dollar value of his claim very substantially by 1933 or '34.

KITCH: Now, of course, index contracts have generally been upheld, and the only problem we have been addressing here is the question of whether some future legislative prohibition of them would be upheld. The principal paper, in discussing index contracts, doesn't explicitly say that it is talking about indexing contracts based on the consumer price index, and I assume as a matter of theory we are not.

But I would suggest that, as a matter of history, we may be going through a kind of episode not unlike the *Gold-Clause Cases*. This is the development of highly standardized boiler-plate language that is put into a large number of contracts without any particular thought about its appropriateness for the risk contemplated by the parties, and this may generate a substantial problem for later enforcing courts.

I think what we are talking about in the move toward universal indexing is use of the consumer price index, but that is not a very subtle instrument for adjusting for various changes in currency values in a vast variety of commercial contracts. As the economists can report more accurately than I, it is a very specialized computation that, as I understand it, represents the average cost of a certain package of services and commodities for an urban family with a given income in a certain fixed situation. It does not in fact accurately reflect what inflation is doing to various other persons or purchasers in various other economic situations with different purchasing needs and options.

BRUNNER: It doesn't even reflect what it does to that particular family.

KITCH: Now it seems to me that if you then assume that the lawyers follow through and start inserting a kind of universal form of CPI boilerplate, and it gets inserted in all sorts of contracts that are not really appropriate to the underlying reason for which the CPI is computed, you may then start to get discussion in which the government says, well, we originally got in the business of putting out this CPI because it's a statistical indicator to assist us in making various planning decisions. We didn't think about it in terms of a contractual price adjustment figure. If it begins to perform that function, that's a very different function, and you can't expect us to ignore the function it is performing in that context in our preparation of the index.

And, therefore, for instance, it might say that we're now going to have two indices. We'll keep the old one, but we'll change the name, and we'll make it clear that all contractual clauses that refer to an index are referring to a new index that we will call, let us say, "the contract price escalator index," which we will have prepared scientifically in light of that function.

It seems to me all of that would raise no serious constitutional problem, and it would get us right back to all these problems about frustration and purpose and so on, and would likely be quite easily upheld.

Second, I would suggest on a number of other grounds that the CPI is in fact a very poor device to use, first of all because of the very real possibility of price controls as a device for controlling the CPI. If that's the contingency you're worried about, then the CPI is a poor device contractually.

That raises the question about what else you can use, and I think that there are a whole set of problems there, and perhaps gold clauses appear attractive relative to other poor choices. I suppose you can use other commodity prices. Maybe you could create within the contract a device for the computation of some type of index. I wonder whether you could put in a clause that it was the CPI as it was computed at a particular time and provide for deviation procedure if the government changes its procedures.

I also think that the real haven of safety is to get away from the CPI altogether because I think any legislation is likely to reflect worry about the CPI, and the various other forms of adjustment provisions may be ignored in such legislation. Only the common clause will be adjusted.

On the other hand, I think the discussion yesterday suggesting constitutional protection of indexing and so on really underrates the potential creativity of contractual relations if parties really seriously wish to address themselves to how they're going to protect against inflation's risks. There are all sorts of ways to do that other than overt forms of indexing.

Perhaps the easiest is some sort of price premium based upon the parties' evaluation of the risk, which won't even appear as any kind of changing price provisions in the contract. Take the floating-rate notes of City National, for instance. It is suggested in some of the papers that this shows a concern about indexing, but the real concern there, of course, was the actual price level that the formula resulted in, not the fact of the formula. It was the fact that the price level was higher than the minimum that the savings and loans were permitted to charge in its impact on savings flows, not the fact that the figure could theoretically vary. Of course, even that itself was not indexed on the CPI;

it was based on other market prices.

I think Milton Friedman's suggestion that one of the ways to make sure that indexing is protected is to have government instruments issued which have index clauses is exactly wrong. Once it becomes a common clause, even a clause in government instruments, that will be the justification for abrogating it, just as occurred in the *Gold-Clause Cases*. If it becomes a common and universal clause, that's why it's so important to get rid of it.

MELTZER: I'd like to comment on the public goods problem. There are some corrective aspects to the use of money, which is, I think, the Buchanan-Tideman point, but I don't see where that has very much to do with this issue. As they make very clear in their paper, there are two aspects. One is the question of defining the weight and measure. The other is the question about protecting the real purchasing power over goods. No one, I think, would ever argue on the basis of history or economic theory that the government was a very active agent in protecting the real value. Quite the contrary, governments historically debased the currency. That's why there were marks around the edges of new coins. So, on the question of protecting the real purchasing power, it's very difficult for me to see where a public goods aspect even enters into that discussion. The question is how do you protect the value of stable purchasing power from the government, and it would seem that that has always been the proper question.

MASHAW: It seems to me that there was one point on which Buchanan, Tideman, and Milton Friedman agreed, and that was that, behind the veil, the monetary constitution made predictability its preeminent value. There was also, at least on Friedman's part, a sense that a major criticism was that government was unresponsive. Now, I realize that there are probably ways to make this superficial contradiction appear consistent, but predictability and political responsiveness are not often thought of as companions that travel together, and it seems to me that we should explore what the implications are of those words. What is meant by predictability and what is meant by responsiveness?

NUTTER: Posing the veil of ignorance, the Buchanan-Tideman paper creates what seems to be a really oversimplified conception of what the social agreement would be. As I understand it, predictability emerges, but after a very cautious statement that one thing you might want is predictability, they seem to jump to the fact that the only thing you need concern yourself about is predictability. And that seems to me an enormous jump. I've been cautioned by my economist friends that free goods are very rare. I wonder why it is that predictability has become a free good. And if predictability is not a free good, one would want to know what the price of predictability is, and there must undoubtedly be a price. And how much predictability? If we identify the price and who pays, it seems to me that we may find ourselves in difficulties spelling out any kind of a monetary constitution. I could imagine a world, I guess, with too much predictability. It must be the "right" amount, and that depends upon what it costs, and I don't find that just saying you're in favor of predictability solves it.

BUCHANAN: I'd like to make just several points. One, Jerry Mashaw's point about the difference here between predictability and political response needs to be brought in. If you accept the predictability notion, you do not want political response. They are contradictory things, and my view would be that if you had predictability you would specifically not want the political response that we had talked about yesterday in terms of the Federal Reserve.

208 GOLD CLAUSE CONFERENCE

But now let me answer this general question about the monetary constitution—some of the points that Kirzner made and that Nutter has made here. We were arguing that plausibly there might emerge from behind this veil of uncertainty, or veil of ignorance, these two minimal components of the monetary constitution. Now you might argue, and Israel Kirsner has argued, that in fact these would not emerge. Possibly, but it seems to me plausible to get agreement on these two minimal components of the monetary constitution.

The great virtue to approaching things in this contractarian way is that it allows us conceptually to derive some sort of agreement as to what might emerge amongst individuals. I'm not going to make the John Rawls mistake and say that these would in fact emerge. I merely think you can make an argument that in fact this is what would emerge from a monetary deliberation. But I'm not going to say this necessarily would emerge.

I look on the predictability point similarly to the definition problem. It seems to me that if we can eliminate from our ordinary contractual arrangements uncertainty about what the value of the unit is going to be, if we can make predictions about it, then we've made the whole system much more efficient, and that's the plausible argument I would make. That's really all I want to say on that.

On the other hand, I want to come back to one point that Ed Kitch made and nobody else has made about the *Gold-Clause Cases*. That is, in terms of the government's own obligations the Court did in fact say it was not allowing the government to break its own word and they said it was unconstitutional. It was only a damage question. That's different from the other two cases.

When Henry [Manne] first suggested this topic it seemed a good area in which to get some interaction and some productive discussion between economists and lawyers. I think the discussion here in these two days has confirmed that view. The main thing that worries me is the tremendous growth in government, which simply cannot continue. But how are we going to arrest this? How are we going to turn things around? I think the only hope really we have is by somehow reinstilling in people what I call a constitutional attitude. I've been very happy to hear the lawyers here these two days, but my own feeling is that these are completely unrepresentative lawyers. It seems to me that most law schools, as I've said and been quoted on, are the most subversive institutions in our country today. It seems to me that we're getting generations of young lawyers who will be our judges, with the view that the way to reform the world is simply by judicial legislation without any real limits.

It seems to me that if we can recover this very central attitude, and if we can begin to get a few lawyers thinking about that—and I think some of them here are—I'm very encouraged.

Very few economists approach these problems as I do. Too many of them lay it down normatively. We draw a social welfare function and then we go out here and maximize to our heart's content. This is why I'm encouraged by what I call the contractarian revival. Many people don't like John Rawls and I think John Rawls made a mistake in being too specific, but I'm very encouraged by the reaction that his book has got among economists.

The great virtue of Rawls' book is that it jumps out of the normative framework and asks what can we derive on contractarian grounds. He goes too far, but at least the process is what should be emphasized, this contractarian constitutional approach to the question, not only economic policy but policy generally. It is something that the lawyers can contribute a great deal to, and I would hope economists would begin to get involved a little bit more in this interaction, into this continuing discourse.

Index

Abortion, access to, and government
    regulation, 97, 98, 101
Adams, William H., 131
Agricultural Adjustment Act, 32–33,
    138
Agricultural prices, and gold price,
    72
Alchian, Armen, 131, 163–64, 166
American Bankers Association, 172
American Writing Paper case, 108
American Writing Paper, Holyoke
    Water Power Co. v., 108n
Ames, Champion v., 101n
Anderson, Martin, 131
Anti-Inflation Act of 1975
    (proposal), 4, 110, 114, 118
Anti-inflation sentiment, 136
Automatic vs. managed monetary
    systems, 48–53, 88, 148, 154,
    167
Ayau, Manuel, 131

Balance of payments, monetary

theory, 137, 138
Baltimore and Ohio Railroad, 3
Baltimore and Ohio Railroad Co.,
    Norman v., 3, 28n, 52n, 93n,
    144
Bank failures, 22–23, 73, 139,
    150–51
Bank Holding Company Act, 116
Bank holiday (1933), 23, 74
Bank of Nova Scotia, 171
Bank of the United States, failure,
    73, 74n, 150
Bank Secrecy Act, 173
Bankruptcy, 19, 96, 163
Basket of currency, 179, 180
Baumol, William J., 53n
Bimetallic controversy, 84, 160
Black, C., 101n
Bond contract, 179
Boulware, Lemuel, 131
Briddle and Mitchell, United States
    v., 59n
Bronson v. Rhodes, 14n, 160

211

212   INDEX

Brunner, Karl, 131, 151, 165, 168,
    176, 204
Bryan, William Jennings, 160, 188
Buchanan, James M., 68n, 119,
    125–29, 131, 138, 139, 148,
    149, 162–63, 185–86, 207–09
Buchanan, James M., and Tideman,
    T. Nicholas, 4, 9–69, 76, 83ff,
    94ff, 107n, 108, 147, 151, 155,
    176, 198, 201, 206, 207
Buchanan, James M., and Tullock,
    Gordon, 42n
Buckley, James L., 62
Burns, Arthur, 60n, 68n, 113

Carothers, Neil, 158
Cassell, Gustav, 26n
Center for Studies in Law and
    Economics, 1
Central banks, 139, 162
    and monetary system, 86–87
    and U.S. gold circulation, 26
Champion v. Ames, 101n
Chandler, Lester, 131, 153–58, 163
Circulation of gold, U.S., 122
    1930s, 32, 77
    1920s, 26
Citicorp, 68n, 81
Clum, Dennis P., 131
Coins, numismatic, 6, 170, 174, 175
Commercial banking system, 21
Committee for the Nation, 156, 159
Commodity standard, in monetary
    system, 50, 51
Commodity use, for money, 85, 86,
    120
Confidence, fluctuation, in monetary
    systems, 13, 16, 21, 22, 27, 31,
    72, 74, 121, 122, 136, 200
Confiscation of gold, purpose, 177
Confiscation of private gold (1933),
    3, 32, 33, 37, 76, 94, 145, 147,
    176, 195
Congress, U.S., 3, 4, 5, 23, 28, 96,
    136, 138, 140, 142, 144

and money value regulation, 29,
    31, 41, 48, 63, 98, 113, 144
Constitution, purpose of, 102
Constitution, U.S., 5, 29, 30, 84, 93,
    183, 193
    and Congressional money regu-
        latory power, 29, 41, 48, 93,
        183
    and contract clause, 183, 184
Constitutional anarchy, 68n
Constitutional attitude, 209
Constitutional change, defined, 28
Constitutional constructionists, strict,
    10
Constitutional interpretations, and
    New Deal monetary powers,
    12, 28–41
Constitutional law, 182, 193
    and gold clause abrogation, 11,
        28–41, 83
Constitutionalism, 182, 183
Consumer price index, 7, 203, 204,
    205
Contract clause, 183, 184, 185, 193
Contract, liberty of, 93, 185
Contract purpose, frustration of
    (doctrine), 5
Contractarian analysis, 4, 55
Contractarian framework, 10–11, 76,
    77, 79, 208, 209
Contracts, private, 27, 49
    and deferred payment, 20
    and government abrogation, 7,
        9–10, 23, 28, 63
    and governmental restrictions,
        53–57, 63ff
    and welfare economics, 12, 53–57
Contractual criteria, and monetary
    constitution, 11, 41–47
Contractual origins, monetary rules,
    41–47, 125ff
Corwin, Edward, 183
Cost-of-living wage contracts, 10, 62
Courts, legislature, power separation,
    5, 30

Credit Anstalt, failure, 73
Credit contraction, and money
  supply, 19, 21, 22
Creditor-debtor relations, 20, 21, 22,
  25, 37, 39, 40, 56, 65, 89, 157,
  164, 197
Currency, basket of, 179, 180
Currency convertibility, 16, 18, 21,
  23, 24, 25, 27, 32, 33, 35, 36,
  37, 55, 56, 73, 75, 76, 121,
  128, 135, 151, 165
  and gold, pre-1933, 13
Currency, demand deposits, issuance,
  17
Currency, deposit, expansion, and
  interest rates, 18, 19
Currency, devaluation, 18, 73
Currency, gold-linked, 16, 80
Customs and Patent Appeals, Court
  of, 187

Dam, Kenneth W., 132, 164, 171
Danbridge v. Williams, 98n
Dauer, Edward, 132
Dawson, John P., 29n
Dawson, John P., and Coultrap, Will,
  52n
Day-Brite Lightening v. Missouri,
  97n
Debs, Richard, 168
Debt contracts, 66
Debtor-creditor relations, 20, 21, 22,
  25, 37, 39, 40, 56, 65, 89, 157,
  164, 197
Deferred payment, 52, 62, 63
  and role of money, 20, 22
Definitional purpose, in monetary
  constitution, 44, 46
Deflation, 32, 40, 67, 72, 90, 95, 164
DeGaulle, Charles, 80
Delegation of monetary authority,
  42ff
  and government failures, 57–61
Demand deposits, currency, 196
  issuance, 17

Demsetz, Harold, 132
Depression, 5, 67, 135
  See also Great Depression
Devaluating, of currencies, 18
Devaluation, 21, 24, 26, 35, 37, 75,
  76, 188
  and monetary base expansion, 33
  as inflationary measure, 39
  of U.S. dollar, 3, 23, 25, 57
  See also Dollar, U.S.
Dickenson, John, 29n
Dietze, Gottfried, 132, 182–84, 185,
  186, 191, 193
Discretionary monetary management,
  failure, 87
Dollar, U.S.
  devaluation, 3, 23, 25, 26, 57, 89,
  135
  and government profits, 33, 34,
  38, 40, 89, 105
  gold link, 24
  purchasing power, 154
  value, 1970s, 61
  value, 1934, 23, 26
  value, pre-1933, 13
Douglas, Paul, 111, 112, 116
Dual money system, 78
Due process clauses, 184, 193
Dunne, Gerald T., 4, 29n, 94n,
  107–18, 119, 132, 140–41,
  142, 148, 150

E. C. Knight Co., United States v.,
  101n
Economic analysis in legal research, 1
Economic history in gold cases, and
  role of gold, 151–68
Economic rights, and government
  regulation, 97, 98
Economic theory, evolution and
  institutional change, 84
Economic vs. individual rights, 98,
  100
Economics, law, relationship, 4, 10,
  207–09

Economists' National Committee on
    Monetary Policy, 158, 159
Effinger v. Kenney, 96n
Employment, and inflation, 64
Employment Act (1946), 110
Employment lay-offs, 19
Epstein, Richard, 132, 174, 178,
    179–80, 189, 190–93, 194,
    195, 196, 201–02
Equilibrium, in international trade,
    15, 16
Escalator clauses, 2, 7, 10, 81
Expectations, and government
    restrictions, on gold dealings,
    clauses, 11, 142
Externalities, in government
    restrictions on private
    contracts, 53ff

Federal Deposit Insurance, 27
Federal Reserve Act, 153
Federal Reserve System, 36, 39, 61n,
    74, 77, 81, 87, 88, 111ff, 139,
    140, 148, 164, 165, 167, 168
    and role of gold, 151
    bank failures, 23
    currency issue, pre-1933, 13
Federalism, 103
Fenno, Veazie Bank v., 32
Fiat money, 49, 50, 55, 166
Fiduciary money, 48, 66, 149
Fisher, Irving, 41n, 84, 158, 159, 160
Fleming, Macklin, 69n, 109
Fractional reserve base, monetary
    issue, 13, 17, 21, 22, 27, 31,
    48, 49n, 54, 120, 138
France, Anatole, 109
Frankfurter, Felix, 31
Franklin National Bank, failure, 73
Free gold problem, 152
Free market, and government
    regulation, 183
Freedom, individual, 49, 98
    and governmental restrictions on

private contract, 53ff
and pre-1933 monetary
    constitution, 13
and private gold trade, 77ff
Freedom of contract, 190, 191, 192
Freedom of expression, 104
Freedom of speech, 103
Freedom of trade, contract, 54, 95ff
Friedman, Milton, 4, 9n, 71–81, 88,
    105, 105n, 109, 109n, 114,
    120, 121, 122, 123, 127, 128,
    129, 132, 135, 137–40, 142,
    143–44, 145–46, 147, 148–49,
    149–50, 151, 152–53, 162,
    166–68, 176, 198, 206
Friedman, Milton, and Schwartz,
    Anna J., 22n, 38n, 47n, 74,
    75, 110, 110n, 121, 138, 151
Frustration, doctrine of, 142, 144

General Accounting Office, 114
Germany, hyperinflation, 79, 136
Glass-Steagall Act, 152, 158–59
Gold, 135
    and U.S. dollar value, 13
    as international money, 80
    as savings media, 170
    circulation, 122
    in U.S., 1930s, 32, 77
        1920s, 26
    confiscation of, purpose, 177
    confiscation of, 1933, 3, 32, 33, 37,
        76, 94, 145, 147, 176, 195
    in Federal Reserve System, 151
    in foreign exchange market, 136
    legal position, after 1974, 169–78
    role of
        and economic history in gold
            cases, 151–68
        in Great Depression, 71–77
    U.S. citizen ownership, 1, 5, 6, 13,
        23, 25, 30, 56
    legalization, 6, 7, 9, 172
    prohibition, 3, 9, 23ff, 30, 32,

37, 56, 58, 60, 128, 148, 173, 174, 175
value of, 3, 5, 6, 23, 27, 33, 36, 75, 121, 145, 150, 161, 170, 173, 180
Gold assets, and tax evasion, 170
Gold bullion, 171, 172
Gold certificates, 76, 161
after 1974, 169, 170, 171
in gold purchases, 38
Gold clause cases, 2, 3, 4, 28, 39, 41, 93, 96, 108, 144, 149, 178, 186, 195, 196, 203, 206, 208
Gold clause legislation, and indexation, 79ff
Gold clauses
abrogation and constitutional law, 11, 28–41, 56, 182–83, 195
abrogation of, 3, 4, 5, 9, 11, 23, 25, 28, 31, 34, 39–40, 63, 64, 65, 67, 80, 83, 89, 93, 94, 128, 136, 141, 142, 157, 165, 181, 183, 187
and indexing provisions, 178–88
historical origins, 143, 160
in contracts, 2, 3, 6, 7, 23, 33, 34, 52, 56, 63, 93, 94, 119, 128, 171, 181
purpose, 95, 185
Gold coin standard, 161
Gold-currency ratio, 139
Gold drain, 18, 23
Gold futures market, 179, 180
Gold-linked currency, 16ff, 80
Gold monetary systems, 14–16
Gold-money ratio, 139
Gold, money, role of, 4, 5
Gold notes, 170, 171
Gold price, and agricultural prices, 72
Gold-related monetary system, U.S., 23
Gold Reserve Act (1934), 23, 173
Gold reserve requirement, 139, 152

Gold reserve system, 21
Gold reserves, 161–62
Gold standard, 86, 120, 121, 137, 148, 168
conceptual base, 14ff
departure from, Great Britain, 13
monetary order, 13
Gold Standard Act (1900), 153
Gold stock, augmentation, 17
Gold trade, private
justification of prohibition, 77–79, 142
prohibition, 23, 25ff, 37, 60, 77
Goldsborough, Alan, 160
Goodrich, Pierre, 181, 197
Government abrogation, of private contracts, 7, 28
Government monetary authority, and inflation, indexation, 61–67
Government, powers, constitutional limitations, 102
Government profits, and dollar devaluation, 33, 34, 38, 40, 89, 105
Government regulation, and free market, 183
Government, state, role in monetary agreements, 11, 28, 89, 94
Governmental authority, monetary, limits, 9–69
Governmental restrictions, on private contracts, 53–57
Great Depression, and role of gold, 71–77
Greenback inflation, 2
Greenbacks, 143
in dual money system, 78

Harberler, Gottfried, 132
Harlan, John Marshall, 108
Harrison, George L., 139, 140, 156
Hart, Henry M., Jr., 29n
Hauser, Rita, 80
Harwood, Edward, 158

Hayek, F. A., 89
Hettinger, Albert J., Jr., 121–22
High-powered money, 75, 122
  examples, 54
Hillendahl, Wesley H., 132
Historical determinism, 10
History, approaches to, 127, 128, 138
Hoarding, of gold, 55, 170
  prohibition, 23
Holyoke Water Power Co. v.
  American Writing Paper, 108n
Holzer, Henry Mark, 12n, 30n, 59n,
  132, 171–76, 180, 195–96
Honnold, John O., Jr., 132
Hoover, Herbert, 150
Hughes, Charles Evans, 41n, 52n,
  142, 144
Hundred Days precedents, by
  Roosevelt, 187, 188
Hyperinflation, in Germany, 79, 136

Income contraction, and public
  confidence, 74
Income, monetary assets, ratio, 14,
  15
Income taxes, and indexation, 10, 62
Index clause
  and constitutional protection, 106,
    120
  bonds, notes, 10
  in contracts, 63, 63n, 96
Index contracts, 4, 6, 10, 41, 67, 80,
  81, 203
Indexation
  and gold clause legislation, 79ff,
    83–84
  and transfer of wealth, 91–92
  government opposition to, 68n, 92
  inflation, and government mone-
    tary authority, 61–67, 90
  prohibition, 67
Indexing, 2, 4, 12, 95, 97, 108, 109,
  203, 205
Indexing provisions, and gold
  clauses, 178–88

Individual evaluation of monetary
  arrangements, 43ff
Individual rights, and government
  regulation, 97, 98
Individual vs. economic rights, 98,
  100
Inflation, 2, 4, 5, 6, 7, 10, 12, 32, 35,
  36, 51, 120, 166, 168, 181,
  182, 185, 197
  as taxation, 105
  control, failure, 10
  greenback inflation, 2
  indexation, and government
    monetary authority, 61–67, 90
Instability of credit, 87
  and gold standard development,
    86
Interest rate, credit, 153
Interest rates, and currency, demand
  expansion, 18, 19
Internal Revenue Service, 170
International liquidity crisis, 59
International trade, equilibrium in,
  15, 16
Interventionist judicial approach, 98
Inventories, and money supply, 20

Jackson, Robert H., 109
Johnson, Harry G., 4, 83–92, 120,
  125, 126, 132, 135–36, 143,
  146, 153, 170, 176–77, 180
Jones, Ronald W., 132, 169
Judicial function, change, 108, 209

Keeler, Davis E., 132, 170–71, 179
Kemmerer, Donald, 132, 158–62,
  163, 178, 188–89
Kemmerer, E. W., 158
Kennedy, Joseph P., 140
Kenner, B. Nowlin, 41n, 47n, 127
Kenney, Effinger v., 96n
Keynes, J. M., 108n
Kirzner, Israel M., 132, 198–201, 208
Kitch, Edmund W., 133, 186–88,
  197–98, 203–04, 204–06, 208

Korzybski, Alfred, 109, 109n, 111

Law and money, interface, 9, 10
Law, economics, relationship, 4, 10, 207–09
Lay-offs, of workers, 19
Leff, Arthur A., 133
Legal issues in economics, 1
Legal limits, governmental monetary authority, 9–69
Legal positivism, 10
Legal tender acts, 189
Legal tender cases, 94
Legislature, courts, power separation, 5, 30
Legislatures, and money value regulation, 29
Levine, Michael, 133, 152, 175, 178, 194–95, 197
Libertarian anarchists, 45n
Liberty Fund, 181
Liberty of contract, 93, 185
Loans of banks, and gold reserves, 18, 19
London Economic Conference (1933), 156
Low-powered money, examples, 54

McLeod, A. Neil, 133
McLuhan, Marshall, 109, 109n
Managed vs. automatic monetary systems, 48–53, 88, 148, 154, 167
Mann, F. A., 30n, 56n, 57n, 68n
Manne, Henry G., 133, 141, 142, 143, 144, 145, 147, 148, 149, 150, 166, 167, 169, 171, 177, 186, 189–90, 193–94, 197, 208
Manne, Henry G., and Miller, Roger LeRoy, 1–7
Mansfield, Mike, 62
Marquis, Harold L., 133
Marshall, Alfred, 2, 84
Marshall, John, 183–84
Mashaw, Jerry L., 133, 178, 206–07

Meigs, James, 115
Meiselman, David, 133, 169, 170, 196, 197
Meltzer, Allan H., 133, 151, 165–66, 168, 176, 179, 180, 181, 189, 190, 206
Mentschikoff, Soia, 133
Meta-legislation, in judicial function, 108
Miller, Roger LeRoy, 133, 178–79
Missouri, Day-Brite Lightening v., 97n
Mofsky, James S., 133
Monetarist thesis, 110, 112
Monetary agreements, government, state, role in, 11, 28, 89, 94
Monetary arrangements
  alternate, and social-constitutional contract, 48–53
  evaluation by individual, 43ff
Monetary assets, income, ratio, 14, 15
Monetary authority
  and government failures, 57–61
  delegation, 42ff
  governmental
    and inflation, indexation, 61–67
    limits of, 9–69
Monetary base, expansion and devaluation, 32
Monetary constitution, 4, 30, 89, 109, 110, 116, 188–209
  and contractual criteria, 11, 41–47
  before 1933, 13ff
  New Deal, 27
  See also New Deal monetary policy
Monetary growth rule, 88, 129, 138
Monetary issue, fractional reserve base, 13, 17, 21, 22, 27, 31, 48, 49n, 54, 120, 138
Monetary management, rules vs. discretion, 88
Monetary powers, New Deal, and constitutional interpretations,

12, 28–41
Monetary rules, contractual origins, 41–47, 125ff
Monetary structure, U.S., post-1933 modifications, 9ff
Monetary system
and central bank, 86–87
before 1933, historical perspective, 11, 12–22
managed vs. automatic, 48–53, 88, 148, 154, 167
Monetary theory, balance of payments, 137, 138
Monetary unit use, in inflation, 62ff
Monetary, weights, measures, regulation, compared, 41, 199, 200, 201
Money
and purchasing power stability, 84–85
role of, and deferred payment, 20, 22
supply
and credit contraction, 19, 21, 22
and gold standard rules, 75
and wage, price, expectations, 19, 20
value
concepts, 153–54
regulation, and U.S. government, 47, 63–64
Morgan, J. P., 74n, 150
Multiple currency clauses, 171, 179

National Banking Act, 84
National Monetary Association, 159, 160
New Deal
and constitutional law, 28–41
monetary policy, 11, 12, 14, 22–27, 57, 72, 105
failure, 58
opposition to, 31
purposes of, 31ff

New York Stock Exchange, 6
Nixon, Richard M., 113
Nominalism (principle), 30n
Norman v. Baltimore and Ohio Railroad Co., 3, 28n, 52n, 93n, 144
Nortz v. United States, 28n
Nussbaum, A., 29n, 53n, 65n
Nutter, Warren, 133, 166, 193, 194, 207, 208

Open Market Committee, 140

Palyi, M., 35
Paper standard, 155
Patman, Wright, 68n, 81
Payroll tax bases, and indexation, 62
Pennock, J. R., 29n
Pensions, and indexation, 92
Perry v. United States, 93n
Pike and Brouwer v. United States, 59n
Political processes, in monetary systems, 21, 24, 25n, 68, 107, 163, 168, 188
Post, Russell Z., and Willard, Charles H., 29n
Predictability, in monetary constitution, 44–45, 46, 48, 50, 65, 66, 84, 86, 105, 106, 125, 138, 179, 206, 207, 208
Price level equilibrium, 14, 15, 16, 18, 19, 20, 22, 35, 37, 56, 90, 135, 136, 137, 160
Prices, and government regulation, 97, 146, 148
Property rights, and U.S. Constitution, 193, 194, 195
Protective instruments, of value, 10
Proxmire, William, 112
Public borrowing, and indexation, 91
Public goals reasoning, in monetary theory, 199ff, 202, 206
Public perception of money supply, 19, 20, 22, 73, 74, 86

Purchasing power
  need, 32
  over goods, precious metal, 84, 91,
    206

Random-walk hypothesis, 6
Rationing stamps, and money value,
    147
Rawls, John, 43n, 99, 103, 126, 127,
    208, 209
Real bills doctrine, 158, 165
Real value, protection, 10
Reconstruction Finance Commission,
    141, 156
Redistribution, of property, 136, 194
Remonetization, 176, 178
Reuss, Henry, 112
Reynolds, Alan, 133
Reynolds v. Sims, 108
Rhodes, Bronson v., 14n, 160
Rodriguez, San Antonio Ind. School
    District v., 100n
Roe v. Wade, 97n
Rogge, Ben A., 133, 181–82, 183,
    184, 185, 186, 193, 196–97
Roosevelt, Franklin D., 22ff, 72, 75,
    77, 120, 121, 136, 137, 139,
    140, 149, 155, 187, 188, 197
Rosett, Richard N., 133
Royal Bank of Canada, 139
Rules vs. discretion issue, in
    monetary management, 88, 89
Ryan, John, 133

Samuelson, Paul, 108n
San Antonio Ind. School District v.
    Rodriguez, 100n
Sandalow, Terrance, 134, 176
School equalization cases, 100
Schuchman, Philip, 134
Schultz, George, 68n
Schultz, Helen E., 134, 181
Schwartz, Warren F., 134, 162, 163,
    177
Shay, Jerome, 116n

Shay's Rebellion, 189
Shenfield, Arthur, 134
Sherman, John, 116
Silver, price of, 145–46
Silver question, 46–47
Silver standard, 160, 161
Simon, William, 172
Simons, Henry, 88, 110, 114
Sims, Reynolds v., 108
Smoot-Hawley tariff, 73, 165
Snyder, Carl, 139
Snyder, John W., 111
Social-constitutional contract, and
    alternate monetary
    arrangements, 48–53
Social security benefits, and
    indexation, 62, 81, 92
Spahr, Walter E., 158
Sparks, John, 29n
Sprague, O. W. M., 158
Stable Money Association, 159
Stable Money League, 159
Stanford, Henry King, 134
Strong, Benjamin, 128, 154
Strong, James G., 160
Supreme Court, U.S., 32n, 52, 60,
    93ff, 108, 140, 141, 144, 186,
    195
  and New Deal monetary policy,
    29–41, 184
  contrasts, 1930s–1970s, 40, 68, 97
  gold clause cases, 2, 3, 4, 14n, 28,
    31, 39, 41, 184, 198

Tabular standard, 2
Tax brackets, and indexation, 91
Tax evasion, and gold assets, 170
Tayloe, Willard v., 96n
Third-party effects, in restrictions on
    private contracts, 53ff
Thomas Amendment, 155
Tideman, T. Nicolaus, 119–23, 134,
    136–37, 139, 145, 147,
    164–65, 179
Trading with the enemy (statute), 23

Triffin, Robert, 68n
Tullock, Gordon, 9n, 61n, 134, 166, 202–03

Unemployment, 135
United States government
and money-value regulation, 47
constitutional obligations in monetary agreements, 11
United States, Nortz v., 28n
United States, Perry v., 93n
United States, Pike and Brouwer v., 59n
United States v. Briddle and Nutchell, 59n
United States v. E. C. Knight Co., 101n
Untermeyer, Samuel, 150

Value of money, 206
Congressional regulation, 29, 31
Vanderlip, Frank A., 159
Veazie Bank v. Fenno, 32
Viner, Jacob, 75, 75n, 122, 138
Volsky, George, 134

Wade, Roe v., 97n
Wages, incomes
and indexation, 62, 81

and money supply, 19, 20, 35, 37, 56
Warren, Earl, 127
Warren, George, 72, 76, 121, 136, 137, 138, 139, 149, 155, 160, 184
Weber, Warren, 9n
Weights, measures, monetary regulation, compared, 41, 199, 200, 201
Weistart, John, 134, 178
Welfare economics, and private contracts, 12, 53–57
Westerfield, Ray, 158
Wheat, price, 137, 146
Willard v. Tayloe, 96n
Williams, Danbridge v., 98n
Williams, Raburn M., 134
Wilson, Woodrow, 116
Winter, Ralph K., Jr., 4, 93–106, 119, 120, 125, 126, 127, 134, 140, 141, 142, 145, 146, 147, 148, 178, 184–85, 186, 188, 193, 194, 196
Wormser, Rene A., 134

Yeager, L. B., 43n

Zwick, Charles J., 134